Lithium Batteries - Recent Advances and Emerging Topics

Edited by Alberto Berrueta and Alfredo Ursúa

Published in London, United Kingdom

Lithium Batteries - Recent Advances and Emerging Topics
http://dx.doi.org/10.5772/intechopen.97970
Edited by Alberto Berrueta and Alfredo Ursúa

Contributors
Alberto Berrueta, Alfredo Ursúa, Anand Parejiya, Bouchaib Manoun, Fouad Ghamouss, Hassan Hannache, Ilias Belharouak, Iñaki Lalinde, Jones Alami, Joseba Arza, Juan José Valera, Kai Pei, Loubna Hdidou, Marm Dixit, Mouad Dahbi, Nitin Muralitharan, Pablo Sanchis, Rachid Essehli, Ruhul Amin, Shunrui Luo

First published in London, United Kingdom, 2024 by IntechOpen
IntechOpen is the global imprint of INTECHOPEN LIMITED, registered in England and Wales, registration number: 11086078, 167-169 Great Portland Street, London, W1W 5PF, United Kingdom

British Library Cataloguing-in-Publication Data
A catalogue record for this book is available from the British Library

Additional hard and PDF copies can be obtained from orders@intechopen.com

Lithium Batteries - Recent Advances and Emerging Topics
Edited by Alberto Berrueta and Alfredo Ursúa
p. cm.
Print ISBN 978-1-80355-474-7
Online ISBN 978-1-80355-475-4
eBook (PDF) ISBN 978-1-80355-476-1

For EU product safety concerns:
IN TECH d.o.o., Prolaz Marije Krucifikse Kozulić 3, 51000 Rijeka, Croatia,
info@intechopen.com or visit our website at intechopen.com.

Meet the editors

Alberto Berrueta received an MSc (Hons) and Ph.D. in Electrical Engineering from the Public University of Navarre (UPNA), Spain, in 2013 and 2017, respectively. In 2013, he joined the Electrical Engineering, Power Electronics and Renewable Energy Research Group (Ingeper) at UPNA, where he is currently an associate professor and a linked researcher at the Research Institute of Smart Cities. His research interests include electrochemical modeling, aging, energy management algorithms, and the safety of lithium-ion batteries. To date, Dr. Berrueta has published more than forty contributions to international journals and conferences, authored two book chapters, and edited one book.

Alfredo Ursúa received a BSc, MSc, and Ph.D. (Hons) in Electrical Engineering from the Public University of Navarre (UPNA), Spain, in 2001, 2004, and 2010, respectively. He is a professor in the Electrical, Electronic and Communications Engineering Department, UPNA. He is also Vice-Dean of the Higher Technical School of Industrial and ICT Engineering. He has participated in more than seventy-five research projects funded by public agencies or in collaboration with the industry and he is co-inventor of seven patents. He has supervised five doctoral theses and published more than 100 scientific papers in international journals and conference proceedings. He is a senior member of the Institute of Electrical and Electronics Engineers (IEEE). His research interests include renewable energies, energy storage, hydrogen technologies, and power electronics.

Contents

Preface

In recent years, we have been immersed in technological development, where lithium batteries power our wearable devices, propel electric mobility, and support the integration of renewable energies into our electrical grid. This book is a compendium of the most recent advances in various technical aspects critically relevant to the growing applicability of lithium batteries.

As editors, both of us hail from the Public University of Navarre, a research and educational center located in northern Spain. For decades, we have been researching the applicability of various energy storage technologies in electrical systems, especially related to renewable energies. The current focus on lithium batteries as a central object of our research reflects the increasing interest generated by this technology both among the scientific community and industry.

This book is the result of a fruitful collaboration that transcends borders, bringing together experts from the United States, China, Morocco, and Spain. The four technical chapters in this book are a blend of global knowledge, each dedicated to a technically significant aspect of the development of lithium batteries.

The book begins with an analysis of the challenges and opportunities presented by high-capacity transition metal oxides as anodic materials in lithium-ion batteries. Subsequently, the book addresses thermal runaway, an inherent risk in lithium batteries. The proper knowledge of this phenomenon allows for safe design of systems including a Li-ion battery. The book moves on to explore the technical requirements and design limitations for batteries in aerial vehicles. This is an emerging application of lithium batteries that can lead to a transformation of our daily lives. Finally, the book highlights the latest advances in copper current collectors designed for Li metal batteries, an incipient technology that can revolutionize the field.

This book is a useful resource for engineers, researchers, and other interested readers. If you are an engineer in a company, you will find ideas to enhance the design of your products. If you are starting a research project on lithium batteries, these chapters will bring you up to speed on crucial topics. If you are a senior researcher, each chapter offers valuable information on relevant topics that will enrich your research. This book not only informs but also explores the energy future driven by lithium batteries. Welcome to this journey!

Alberto Berrueta and Alfredo Ursúa
Institute of Smart Cities,
Department of Electrical, Electronic and Communications Engineering,
Public University of Navarre (UPNA),
Pamplona, Spain

Chapter 1

Introductory Chapter: Frontiers of Lithium Battery Technology – Challenges, Opportunities, and Future Horizons

Alberto Berrueta and Alfredo Ursúa

1. Introduction

Lithium-ion batteries (LIBs) stand as the cornerstone of modern energy storage solutions, driving applications such as electric vehicles or grid-scale energy storage. The relentless pursuit of higher energy density, enhanced safety, and increased longevity has fueled an evolving landscape of research and innovation in this field.

This introductory chapter aims to provide a snapshot of the diverse themes explored within the book, "Lithium Batteries – Recent Advances and Emerging Topics." It encapsulates the essence of some of the challenges and recent breakthroughs, concerning lithium battery research.

The history of lithium-ion batteries traces back to the late twentieth century when the pioneering work of M. Stanley Whittingham [1], John B. Goodenough [2], and Akira Yoshino [3] (Nobel Prize in Chemistry 2019) laid the groundwork for this revolutionary energy storage technology. The early breakthroughs in the 1970s and 1980s led to the development of the first commercial lithium-ion battery in the early 1990s for consumer electronics.

The significance of lithium-ion batteries surged further with their application in the automotive industry. The emergence of electric vehicles (EVs) as a sustainable and clean alternative to traditional internal combustion engine vehicles owes a significant debt to lithium-ion technology. These batteries enable the long-range driving capabilities and performance necessary for the widespread adoption of EVs.

However, the historical evolution of lithium-ion batteries also encompasses challenges. Issues of safety, such as thermal runaway and the potential for battery fires, have been notable concerns [4]. Additionally, the reliance on limited and sometimes geopolitically sensitive resources for lithium-ion battery production, such as cobalt and lithium, has raised questions about sustainability and supply chain resilience.

Nonetheless, the importance and impact of lithium-ion batteries cannot be overstated. They not only revolutionized the electronics industry but also played a pivotal role in the transition toward renewable energy and the electrification of transportation. The development of grid-scale energy storage systems also leans heavily on advances in lithium-ion technology to support the integration of renewable energy sources into the power grid [5].

IntechOpen

The evolution of LIBs is a dynamic interplay between overcoming challenges and exploring new opportunities. The chapters within this book encapsulate some of these challenges and pave the way for novel opportunities, showcasing the landscape of LIB technology.

The challenges analyzed in this book, ranging from safety concerns to specific application demands, underscore the evolving nature of the field. Yet, within these challenges lie opportunities. Each hurdle becomes a catalyst for innovation, driving researchers and practitioners to explore new frontiers and redefine the possibilities of energy storage.

As we look to the future, the chapters outlined in this book present a roadmap for innovation. From high-capacity anodes to strategies addressing thermal runaway and dendrite growth, each topic contributes to the broader narrative of LIB advancements. The collaborative nature of these efforts underscores the interconnectedness of researchers, industry experts, and innovators, all working together to overcome challenges and push the boundaries of what LIBs can achieve.

In conclusion, this introductory chapter sets the stage for the deeper exploration of lithium batteries presented in the subsequent chapters of the book. It serves as a call to action, encouraging continued collaboration, exploration, and ingenuity. The journey ahead promises not only to address the challenges that currently define lithium battery technology but also to unearth novel opportunities, laying the foundation for a future where lithium batteries are at the forefront of sustainable, efficient, and groundbreaking energy solutions. As we navigate forward, the chapters that follow will contribute to this narrative, propelling lithium battery technologies toward new horizons and unparalleled excellence.

Author details

Alberto Berrueta* and Alfredo Ursúa
Institute of Smart Cities, Department of Electrical, Electronic and Communications Engineering, Public University of Navarre (UPNA), Campus de Arrosadia, Pamplona, Spain

*Address all correspondence to: alberto.berrueta@unavarra.es

References

[1] Whittingham MS. Electrical energy storage and intercalation chemistry. Science. 1976;**192**(4244):1126-1127. DOI: 10.1126/science.192.4244.1126

[2] Mitzushima K, Jones PC, Wiseman PJ, Goodenough JB. Li_xCoO_2 ($0<x\leq1$): A new cathode material for batteries of high energy density. Materials Research Bulletin. 1980;**15**:783. DOI: 10.1016/0025-5408(80)90012-4

[3] Yoshino A et al. USP4,668,595 and JP1989293, Filing Date (Priority) May 10, 1985

[4] Yang Q, Xu C, Geng M, Meng H. A review on models to prevent and control lithium-ion battery failures: From diagnostic and prognostic modeling to systematic risk analysis. Journal of Energy Storage. 2023;**74**. DOI: 10.1016/j.est.2023.109230

[5] Millet L, Berrueta A, Bruch M, Reiners N, Vetter M. Extensive analysis of photovoltaic battery self-consumption: Evaluation through an innovative district case-study. Applied Physics Reviews. 2019;**6**:021301. DOI: 10.1063/1.5049665

Chapter 2

Challenges and Strategies of High-Capacity Transition Metal Oxides as Anodes for Lithium-Ion Batteries (LIBs)

Loubna Hdidou, Fouad Ghamouss, Bouchaib Manoun, Hassan Hannache, Jones Alami and Mouad Dahbi

Abstract

To satisfy the growing demand for high-energy and high-power-densities Lithium-ion Batteries (LIBs), the design and development of efficient electrode materials are necessary. In comparison to graphite, transition metal oxides (TMOs) have recently been widely investigated as anode materials due to their promising properties. These combine high specific capacities and high working potential, making them attractive anode candidates for emergent applications. Unfortunately, because of their poor electronic conductivity and high-volume expansion during cycling, they are unpractical and difficult to employ. To overcome these limitations, different approaches have been adopted. Examples are synthesizing the metal oxides at the nanometric scale, designing three-dimensional or hollow structures, coating the material with carbonaceous materials, etc. In this chapter, we report the elaboration of nanostructured transition metal oxides (Co_3O_4, Mn_3O_4, $Co_{3-x}Mn_xO_4$) using alginate gelling synthesis method. The Co_3O_4 octahedral-like nanoparticles display a remarkable cycling performance and good rate capability of 1194 mAh g^{-1} at C/5 and 937 mAh g^{-1} at 2C. Partially substituting the Co with Mn was shown to result in the production of $Co_{2.53}Mn_{0.47}O_4$ and $MnCo_2O_4$ with high initial specific discharge capacities of 1228/921 and 1290/954 mAh g^{-1}, respectively. As a Co-free material, the Mn_3O_4 delivers a reversible capacity of 271 mAh g^{-1}, after 100 cycles.

Keywords: lithium-ion batteries, transition metal oxides, high-capacity, anode materials, alginate gelling method

1. Introduction

Since their commercialization, (LIBs) have been regarded as the most efficient energy storage system. They have been integrated into portable electronics and electrical devices. Their energy density has increased gradually until ~250Wh kg^{-1} [1]. The LIBs operating mechanism is based on the lithium-ion exchange between the cathode and the anode materials. During the charging process, an external current is applied to the cell, forcing the lithium-ion to leave the cathode moving to the anode. Inversely, during the discharge, which is a spontaneous reaction, the lithium-ion is transferred from the anode to the cathode. To preserve electrical neutrality, electrons are also transported to the cathode through an external circuit. Then, the electrical energy that was stored as chemical energy during the charging process is produced. The conversion of the chemical energy to electrical energy and vice-versa is based on the redox reaction of metallic elements present at the electrodes. These reactions are controlled by the chemical potential difference. The LIBs performances depend mainly on the combination of their different constituents, including the positive and negative electrodes, the electrolyte, and the separator.

The anode material is one of the critical components of the LIBs and depending on the Li^+ ion reaction with the active material, three different storing mechanisms could be listed. These are the intercalation, the alloying, and the conversion. Graphite is considered the most employed negative electrode material for LIBs [2]. Its working mechanism is based on the Li^+ intercalation/deintercalation into the graphite interlayer, following the reaction: $6C + xLi^+ + xe^{-1} \leftrightarrow Li_xC_6$ [2]. With a theoretical capacity of 372 mAh g^{-1}, the graphite anode delivers a capacity that is higher than 360 mAh g^{-1} and shows good reaction reversibility. Nonetheless, due to its low capacity, other carbonaceous materials were examined, such as graphene [3], hollow spheres, nanotubes [4], nanofibers [5], porous materials, etc. For graphene and hollow carbon spheres, it was demonstrated that the Li^+ storage took place due to the surface adsorption and the interactions with the carbon functional groups [6, 7].

Non-carbonaceous materials were also tested as anode materials for LIBs. Alloying elements such as Si, Ge, and Sn are considered promising alternatives. Particularly, Si exhibits the highest gravimetric capacity of 4200 mAh g^{-1} [8]. The lithiation mechanism happens in two steps, the first one where the $Li_{13}Si_4$ alloy is formed, and the second consists of the formation of $Li_{22}Si_5$ phase [9]. Ge and Sn also show high initial theoretical capacities of 1600 mAh g^{-1} and 993 mAh g^{-1}, respectively [10, 11].

Metal oxides are a very attractive class of materials, they are regarded as one of the most intriguing alternative graphite materials because of their high specific capacities and high working potential. Recently different oxides, such as Fe_2O_3 [12], MnO [13], NiO [14], Mn_3O_4 [15], $CoFe_2O_4$, $MnFe_2O_4$ and $NiFe_2O_4$ [16, 17] were used as anodes. They are generally known to react with lithium by a conversion reaction, where the metal ions are reduced to metallic form and then oxidized during consecutive discharge/charge operations:

$$\mathrm{M_aO_b} + (\mathrm{b.n})Li^+ + (\mathrm{b.n})\mathrm{e}^- \leftrightarrow aM + bLi_{\mathrm{n}}\mathrm{O} \quad (1)$$

Where M=Mn, Fe, Co, Ni, Cu, Ru, Mo, Cr, W, etc.

The In-situ transmission electron microscopy (TEM) technique demonstrates that the lithiation process of metal oxides occurs first via an intercalation reaction with the formation of metastable and intermediate phases, followed by a conversion reaction [18]. During a discharge process, with a metallic lithium counter electrode, the metallic cations are reduced to their nanometric metallic elements and get surrounded by Li_2O which is chemically more stable than any other metal oxide. However, when lithium-ion is electrochemically extracted from the composite (charge), the metallic species clusters get disintegrated and converted to small, dispersed particles with a high active specific surface area. This increases oxygen contact and catalyzes the formation of the metallic oxides as Li_2O decomposes. The charge process involves the deconversion of the metallic species formed during the discharge process and the decomposition of the inactive lithium oxide Li_2O by cleaving the stable Li-O bonds.

Even if metal oxides exhibit high-capacity values, they have been known to suffer from important volume expansion reaching up to 200% during the lithiation process. This induces the electrode's pulverization and disintegration, resulting in the loss of particle contact, a take-off from the current collector, and the formation/deformation of a solid electrolyte interface (SEI) layer, leading to an acceleration of capacity drop [19–21]. To improve the metal oxides cyclability different approaches have been adopted [22, 23].

2. Engineering approaches for high performances transition metal oxides

2.1 Design of three-dimensional (3D) structures or hierarchical configurations

Three-dimensional (3D) structures or hierarchical configurations are generally built from 0D, 1D, or 2D particles resulting in structures with enhanced porosity and relatively large specific surface areas. Three-dimensional structures have several advantages; they provide more channels for Li^+ and e^- diffusion, a larger electrode/electrolyte contact area, and improved stress resistance to volume change during the Li^+ insertion/extraction [21]. A 3D structure made of vertical nanowires of Co_3O_4 with an improved electrical conductivity was prepared by Wang et al. The Co_3O_4 shows stable cycling and a high coulombic efficiency of around 100%. When assembled in a full cell system with lithium iron phosphate, LFP, cathode, the cell exhibits a capacity of 138 mAh g^{-1} with a capacity retention of 96% after 100 cycles [24]. Kong et al. reported the elaboration of 3D hierarchical $MnCo_2O_4$ porous dumbbells exhibiting a stable reversible capacity of 955 mAh g^{-1} after 180 cycles at 200 mA g^{-1} [25]. Hierarchical conversion metal oxides have also stimulated efforts for the development of binder-free negative electrodes for the next-generation LIBs. In a recent study, Wu et al. have reported the elaboration of self-supporting multilevel-3D porous NiO nanowires as a binder-free anode material. The synthesized NiO displayed superior electrochemical performances, with a specific capacity that reached 1816.3 mAh g^{-1} at the current density of 100 mA g^{-1}, and remained at 1318.7 mAh g^{-1} after 150 cycles [26].

2.2 Design of nanoparticles

As an anode for lithium-ion batteries, nanoparticles display superior properties compared to their bulk counterparts. Their reduced size significantly increases the Li-ion diffusion rate. We recall that the characteristic diffusion time constant is given by the equation:

$$\tau_{eq} = L^2 / D \quad (2)$$

where L is the diffusion length and D is the diffusion constant.

Recognizing that the mean diffusion time (τ_{eq}) is proportional to the square of the diffusion length (L), it is safe to claim that when the particle size is reduced, the intercalation time decreases. The nanoparticles' high specific surface area provides more electroactive sites and displays an important contact area with the electrolyte [20]. Recently, a variety of nano-sized metal oxides such as Mn_2O_3 [27], Co_3O_4 [28], Fe_3O_4 [29], etc. have been reported.

2.3 Design of hollow structures

A hollow structure is another alternative to enhance electrochemical properties. It is attracting increasing attention because of its ability to buffer the volume expansion during the cycling process and to increase the contact between the inner active material and the electrolyte due to the presence of voids inside these structures. Jian et al. have prepared Mn_3O_4 hollow spheres, using an aerosol-based "droplet to particle" method and evaluated them as anode material in LIBs. The particles displayed capacity retention of ~980 mA g^{-1} for over 140 cycles at 200 mA g^{-1} [30]. Hollow $CoFe_2O_4$@Fe_3O_4 nanospheres have also shown good cycling stability, after 500 cycles, with a specific capacity of 365 mAh g^{-1} at a current density of 1000 mA g^{-1} [31]. Xie et al. have reported the elaboration of hollow Fe_3O_4/rGO composites with a reversible capacity of 827.3 mAh g^{-1} after 550 cycles at 0.5 A g^{-1} [32].

2.4 Carbon coating and hybridization with carbonaceous materials

To enhance the electronic conductivity and the reaction kinetics during cycling, metal oxide particles are coated with carbon. Alternatively, they are inserted into a carbonaceous matrix. In both cases, the strain of volume change during the Li^+ ion insertion/extraction is better accommodated, and particles aggregation is avoided. A carbon shell can be also an effective barrier between the electrode and the electrolyte, which stabilizes the solid electrolyte interface and prevents electrolyte decomposition. Different studies have synthesized hybrid metal oxide/carbon materials as high-efficiency anodes for advanced lithium-ion batteries.

To produce $CoFe_2O_4$/C fibers, Wu et al. used the electrospinning technique. The $CoFe_2O_4$/C composite was evaluated for its performance as an anode, demonstrating a reversible capacity of 490 mAh g^{-1} after 700 cycles [33]. In a different study, Li et al. reported on the elaboration of $CoFe_2O_4$ nanoparticles, uniformly dispersed, on graphene sheets. This three-dimensional (3D) network exhibited a specific reversible capacity of up to 938 mAh g^{-1}, at a current density of 1000 mA g^{-1} [34]. Nitrogen-doped porous Carbon/Co_3O_4 nanocomposites were obtained via a hydrothermal method using natural porous crawfish shells as the carbon source. This composite delivered an irreversible discharge capacity of 1060 mAh g^{-1} at a current density of 100 mA g^{-1} [35].

In this chapter, we report the synthesis of Co_3O_4, Mn_3O_4, $Co_{2.53}Mn_{0.47}O_4$, and $MnCo_2O_4$ at the nanometric scale for use as anode materials for LIBs, by a bio-inspired simple synthesis method, using alginate biopolymer as a bio-template for crystals nucleation and growth.

3. Design of nanostructured Co_3O_4, Mn_3O_4, $Co_{2.53}Mn_{0.47}O_4$, and $MnCo_2O_4$ as high-performance anode materials for the next generation LIBs

3.1 Nanostructured Co_3O_4 as anode material for LIBs

Following a synthesis procedure by Hdidou et al. [36], Co_3O_4 was prepared via the alginate gelling method. A sodium alginate aqueous solution was prepared under medium stirring at room temperature. Subsequently, hydrogel beads were formed by adding, dropwise, the obtained solution into a homogeneous metal solution containing Co^{2+} cations. The beads were gently stirred in the metal solution overnight before filtration, washing with distilled water, and drying. The sacrificial bio-template was removed by heating the xerogels in airflow at 600°C.

During the synthesis process, the alginate biopolymer could be used as:

- A stabilizer to control the particle size and the specific surface area;
- A template to direct the structure and the morphology of particles;
- An Oxygen donor.

The powder X-ray diffraction (XRD) patterns of the Co-alginate composite obtained after the thermal treatment, confirm the formation of the cubic spinel structure Co_3O_4 with a *Fd-3m* space group (**Figure 1**). The synthesized material's morphology and particle size were characterized by Scanning Electron Microscopy (SEM), as shown in **Figure 2**. The images reveal the presence of polyhedral octahedral-like nanoparticles with a heterogeneous particle size distribution and an average size of 90 nm. The particles exhibit a Brunauer–Emmett–Teller (BET) specific surface area of 12.4 m^2g^{-1}. The morphology and the particle size are relevant parameters that have a direct impact on a material's electrochemical properties. One of the most effective methods to improve electrode kinetics is the elaboration of the active material at the nanometric scale. Generally, nanoparticles provide a huge electrode/electrolyte contact area, a reduced distance for Li^+ diffusion and electron transfer, and more electro-active sites.

To evaluate the electrochemical performance of the nanostructured Co_3O_4 electrode, cyclic voltammetry (CV) and galvanostatic tests were performed. The cyclic voltammetry curve of Co_3O_4 is illustrated in **Figure 3b**. It is seen that during the first scan of Co_3O_4, the main reduction peak is located at 0.66 V and is attributed to the reduction of Co^{3+} and Co^{2+} to Co, and the formation of the solid-electrolyte-interface (SEI) layer. The main oxidation peak was observed at 2.07 V and is associated with the oxidation reaction of Co/Co_3O_4. The proposed Li^+ storage mechanism is according to the following reaction [37]:

$$Co_3O_4 + 8Li^+ + 8e- \leftrightarrow 3Co + 4Li_2O \quad (3)$$

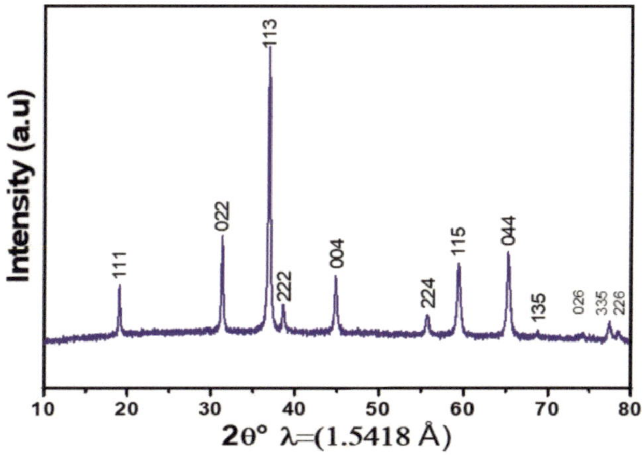

Figure 1.
X-ray powder diffraction patterns of Co_3O_4.

Figure 2.
SEM images of Co_3O_4.

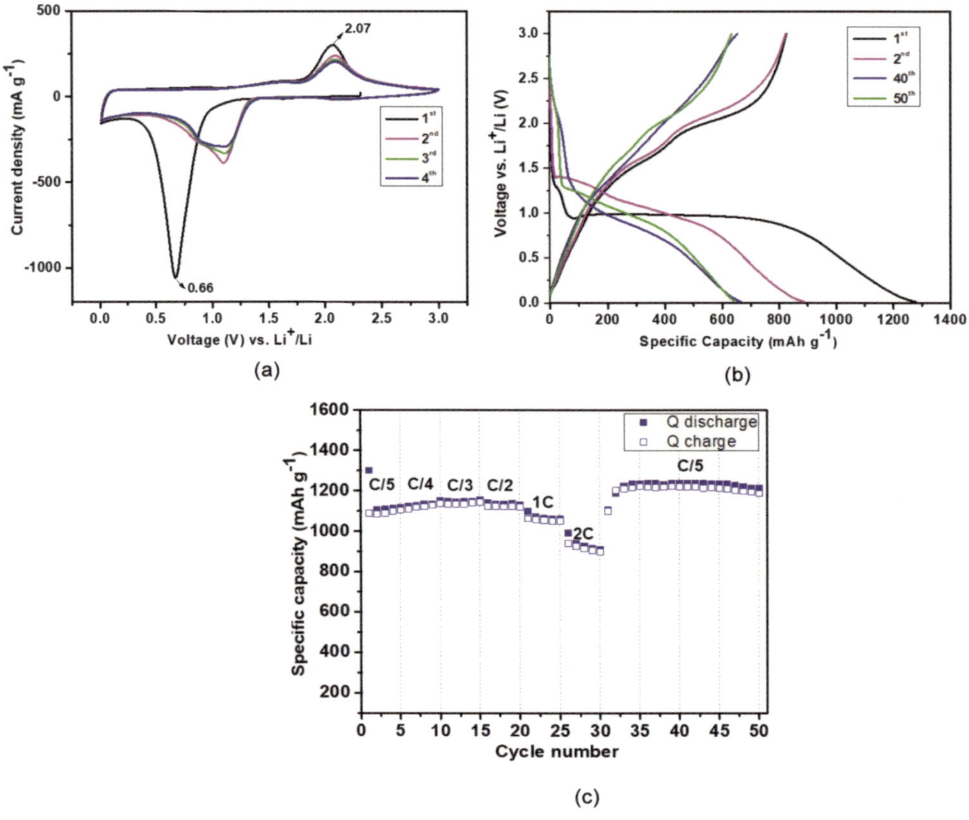

Figure 3.
Electrochemical performances of Co_3O_4 (a) Cyclic voltammogram of the as-prepared Co_3O_4 in the voltage range of 0.01–3.00 V vs. Li/Li^+ at a scan rate of 0.05 mV s^{-1}, (b) Charge-discharge profiles of Co_3O_4 at the current rate of C/5 in the voltage range of 0.01–3.00 V (c) Rate capability of Co_3O_4 electrode material at various current rates from C/5 to 2C.

Further investigation showed that the shape of the CV curve is mostly maintained in the subsequent sweeps, indicating good cycle reversibility of the electrochemical reactions [38, 39]. Starting from the second cycle, the CV curves display overlapped two reduction peaks located at around 1.1 and 0.91 V, assigned to the reversible reduction of Co^{3+} to Co^{2+} and Co^{2+} to Co, respectively. These peaks become broader with lower intensities compared to the first cycle which is due to the pseudo-capacitive behavior of the metal oxide that starts taking place during the following cycles due to the particle size reduction and the irreversible Co and Mn elements formation, indicating the capacity loss. **Figure 3a** displays the galvanostatic charge/discharge performances of the Co_3O_4 electrode, examined at a C/5 current rate in the voltage window of 0.01–3.00 V (1C = 1.23 A g^{-1}). The electrode delivers a high initial specific discharge capacity of 1008 mAh g^{-1} with a coulombic efficiency (CE) of 75%. After 50 cycles the capacity was maintained to 680 mAh g^{-1} with a CE of 99%. During the first discharge curve, a large discharge plateau at 1.06 V appeared, corresponding to the lithium reaction with Co_3O_4 and the formation of SEI. During charging, a plateau was also observed at 1.9 V. This plateau is attributed to a Co/Co_3O_4 oxidation reaction during the lithium-ion extraction process (**Figure 3b**). The enhanced electrochemical performance of Co_3O_4 could be attributed to the reduced particle size of 90 nm and the reactions reversibility during the lithiation/delithiation process. As was expected from the galvanostatic test, Co_3O_4 exhibits good rate capability performances. The cell delivers average specific discharge capacities of 1149, 1133, 1147, 1134, 1070, and 937 mAh g^{-1}, with coulombic efficiencies of 95, 97, 91, 98, 99, and 97% at the current rate of C/5, C/4, C/3, C/2, 1C, and 2C respectively. When the current density was set back to C/5, a high specific discharge capacity of 1222 mAh g^{-1} was recovered with a CE of 96%. Indicating its tolerance to the fast Li^+ lithiation/delithiation process (**Figure 3b**).

3.2 Nanostructured $Co_{3-x}Mn_xO_4$ (x = 0.47, 1) as anode materials for LIBs

The insertion of the Mn into the Co_3O_4 structure was ensured using the alginate gelling synthesis method. The alginate solution was added drop by drop into a homogeneous metal solution containing a mixture of Co^{2+} and Mn^{2+} cations. The obtained gel was then washed, dried, and calcined under an air atmosphere. The substitution of Co^{3+} (0.54 Å) by Mn^{3+} (0.58 Å) cations that have a larger ionic radius was confirmed by the X-ray diffraction (**Figure 4**) and the atomic absorption spectroscopy analysis. $Co_{2.53}Mn_{0.47}O_4$ and $MnCo_2O_4$, with cubic spinel structures adopting *Fd-3m* as a space group, have been produced. Analysis of the SEM images showed that $Co_{3-x}Mn_xO_4$ (x = 0.47, 1) crystallizes into an octahedral-like morphology since the growth of nanoparticles is privileged along the axis <100>. The average particle size calculated by ImageJ software of $Co_{2.53}Mn_{0.47}O_4$ and $MnCo_2O_4$ is 72 and 69 nm respectively (**Figure 5**). Using a scanning electron microscope, coupled with an energy-dispersive X-ray spectroscope (EDS), the elemental composition revealed the presence of Co and Mn as the major elements. This is shown in **Figure 6** where other elements such as Na, and Ca could also be detected.

The mixed oxides $Co_{3-x}Mn_xO_4$ (x = 0.47, 1), demonstrate a promising electrochemical activity with high specific discharge capacities of 1228 and 1290 mAh g^{-1} and an initial CE of 75 and 74% (**Figure 7a** and **b**). This capacity could not be maintained in long-term cycling, which was also observed by Lai et al., who reported that with increasing the cycle numbers the Mn element gets dissolved into the electrolyte and settles down on the electrode surface [40]. From the cyclic voltammetry profile, the two cathodic peaks were observed during the conversion reaction for each compound (1.21 V/0.70 V, and 1.28 V/0.62 V), they are assigned to the reduction process of Mn/Mn^{3+} and Co/Co^{3+}, respectively. However, three of the observed anodic peaks, 2.09/1.52/1.10 V, 2.01/1.52/1.13 V, could be attributed to the oxidation of Co to Co^{2+} and Co^{3+}, and Mn to Mn^{2+} [41] (**Figure 7c** and **d**).

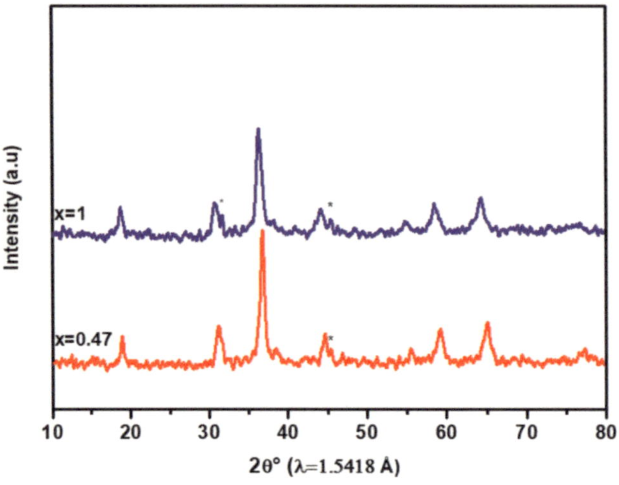

Figure 4.
X-ray powder diffraction patterns of $Co_{2.53}Mn_{0.47}O_4$ *and* Co_2MnO_4.

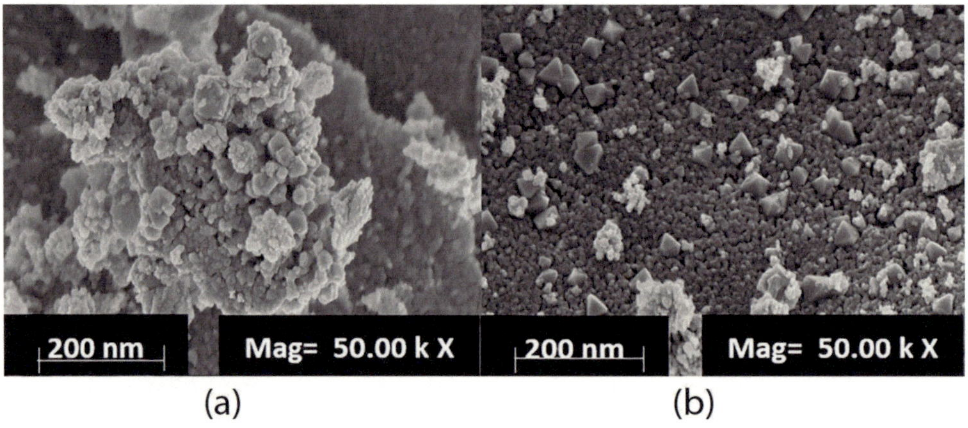

Figure 5.
SEM images of the synthesized oxides (a) $Co_{2.53}Mn_{0.47}O_4$ *and (b)* Co_2MnO_4.

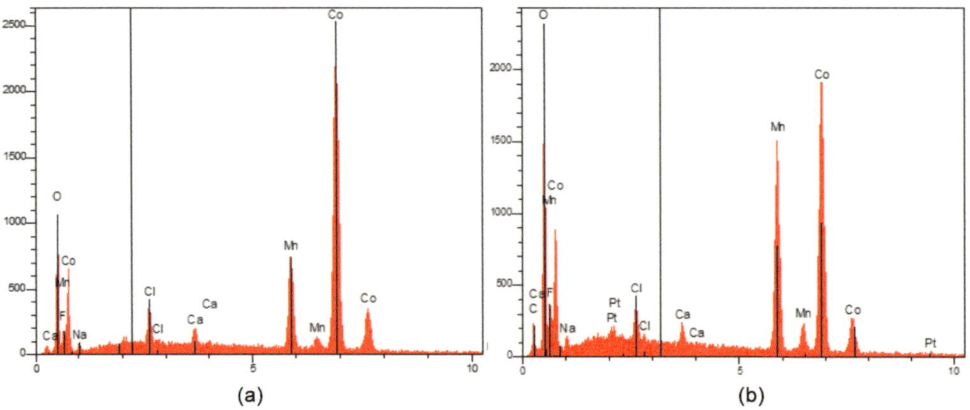

Figure 6.
EDS analysis of the synthesized oxides (a) $Co_{2.53}Mn_{0.47}O_4$ *and (b)* Co_2MnO_4.

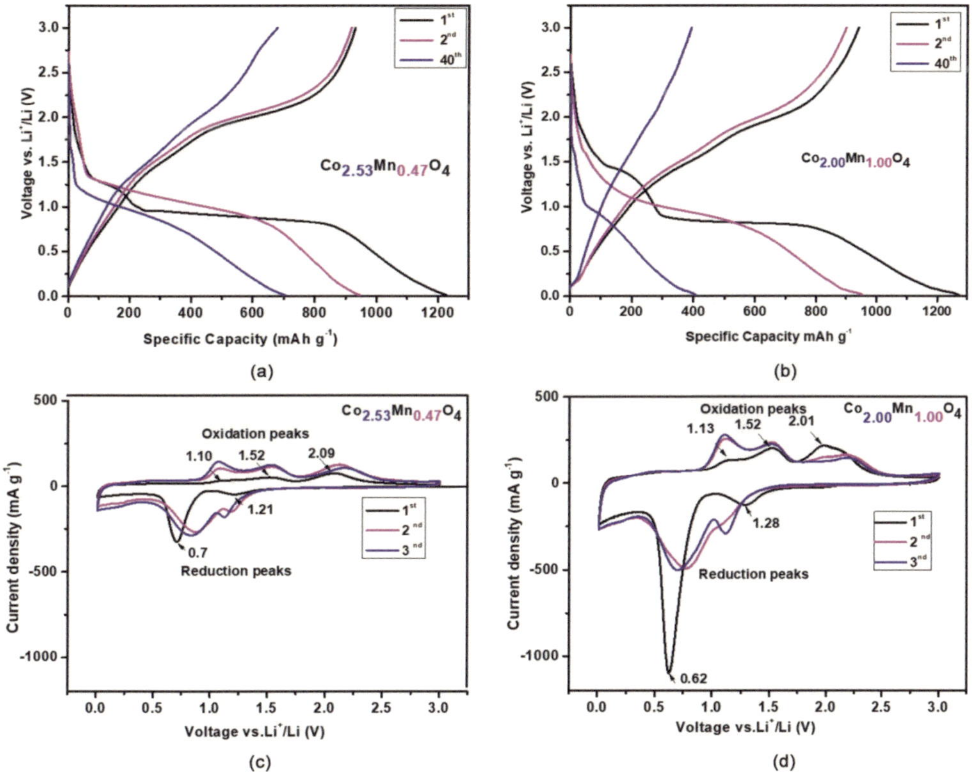

Figure 7.
Electrochemical performances of $Co_{2.53}Mn_{0.47}O_4$ and Co_2MnO_4 (a, b) Charge-discharge profiles of $Co_{2.53}Mn_{0.47}O_4$ and Co_2MnO_4 at the current rate of C/5 in the voltage range of 0.01–3.00 V (c, d) Cyclic voltammograms of the as-prepared $Co_{2.53}Mn_{0.47}O_4$ and Co_2MnO_4 in the voltage range of 0.01–3.00 V vs. Li/Li^+ at a scan rate of 0.05 $mV\ s^{-1}$.

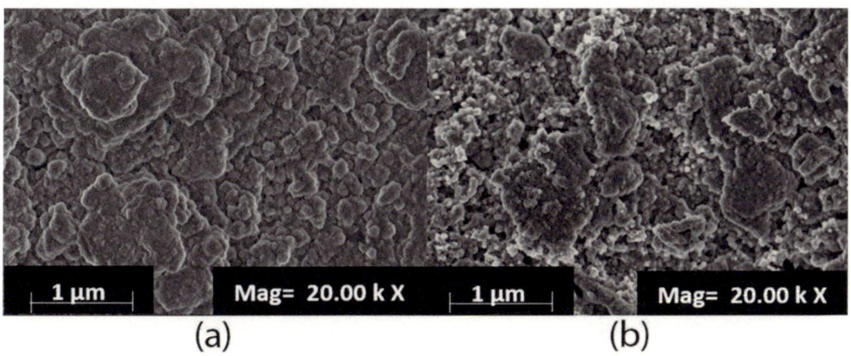

Figure 8.
SEM images of the electrodes (a) $Co_{2.53}Mn_{0.47}O_4$, and (b) Co_2MnO_4, after 2 cycles.

The particle morphology was also examined. **Figure 8a** and **b** revealed that after two cycles, the nanometric size distribution was retained and the structure was still observed. Nonetheless, a thin layer was formed on the electrode surface. It refers to the SEI films that originated from the decomposition of the electrolyte and the consumption of salt, and solvent species [42].

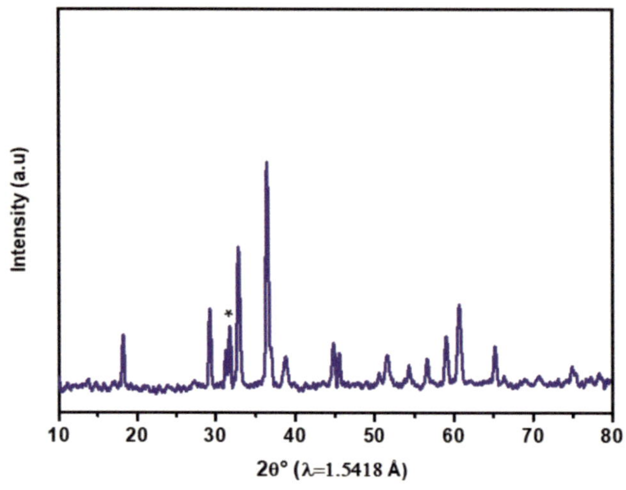

Figure 9.
X-ray powder diffraction patterns of Mn_3O_4.

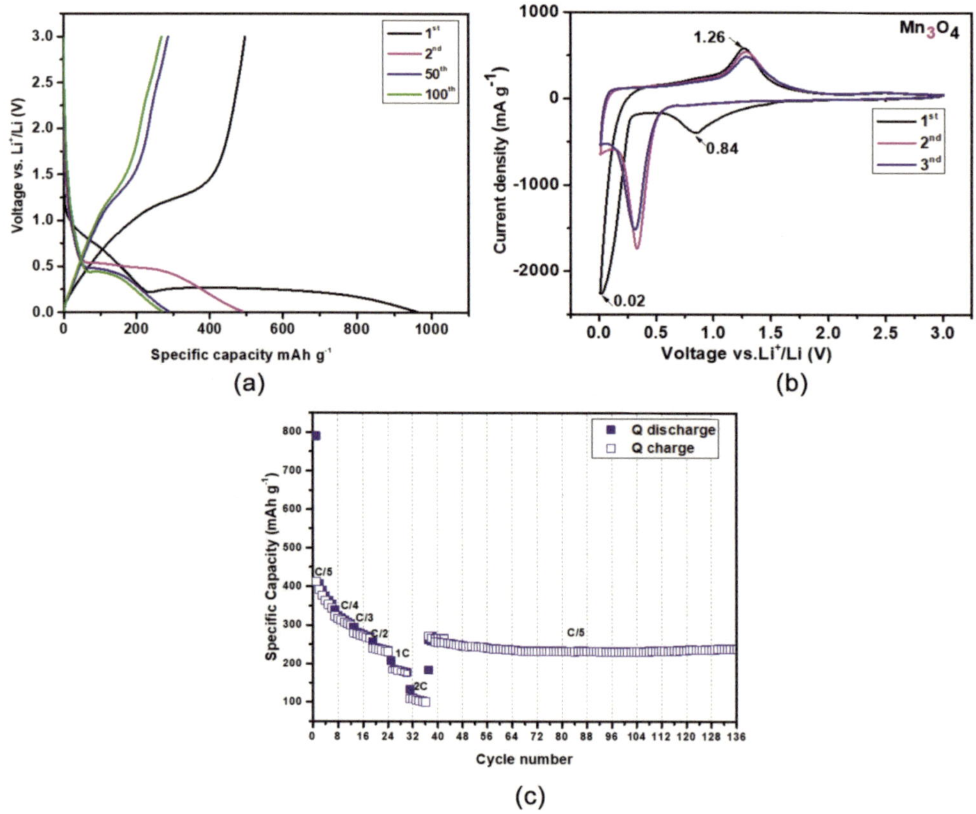

Figure 10.
Electrochemical performances of Mn_3O_4 (a) Charge-discharge profiles of Mn_3O_4 at the current rate of C/5 in the voltage range of 0.01–3.00 V, (b) Cyclic voltammogram of the as-prepared Mn_3O_4 in the voltage range of 0.01–3.00 V vs. Li/Li^+ at a scan rate of 0.05 $mV\,s^{-1}$, (c) Rate capability of Mn_3O_4 electrodes material at various current rates from C/5 to 2C.

3.3 Nanostructured Mn_3O_4 as anode material for LIBs

The elaboration of Mn_3O_4 by the alginate gelling method is also reported in this chapter. The Mn_3O_4 spinel oxide was produced by calcining the hybrid Mn-alginate xerogel at 600 °C under the air atmosphere. This phase adopts *I41/amd* as a space group. Mn_3O_4 is a tetragonal distorted spinel with an ionic structure of $Mn^{2+}\left[Mn_2^{3+}\right]O_4$, where Mn^{2+} are located in the tetrahedral sites and Mn^{3+} in the octahedral sites. The sample shows an impurity peak which was indexed as Mn_5O_8 (**Figure 9**).

As a free Co material, Mn_3O_4 shows moderate electrochemical performance. The initial specific discharge/charge capacities delivered by the prepared material are 960, 497 mAh g^{-1} with a CE of 51%. Between the first and the second cycle, the discharge capacity drops to its half. Nevertheless, between the 50th and 100th cycles, only a loss of 14 mAh g^{-1} was observed and a capacity of 271 mAh g^{-1} was maintained (**Figure 10a**). With increasing the cycle number, the specific capacity gets more stable. This stability could be attributed to the activation process resulting from the increase of the lithium-ion reactivity with the appearance of Mn^{2+}/Mn^{4+} redox reaction [25]. The coulombic efficiency rises to 98.9% indicating its good capacity retention. At the first discharge curve, a pseudo-plateau at 0.70 V with a large plateau at 0.24 V took place, which is related to the lithium insertion reactions and the electrolyte decomposition (**Figure 10a**). From the CV test, in the first cathodic scan, the broad peak at 0.30 V is attributed to the reduction of Mn^{3+} into Mn^{2+} and the formation of a solid-electrolyte-interface (SEI) layer (peak around 0.84 V). The intensive sharp peak in the low potential region is attributed to the reduction of Mn^{2+} to Mn leading to the formation of Li_2O, this peak moves to 0.32 V in other cycles. During the first charge process, the broad anodic peak at 1.27 V can be ascribed to the simultaneous oxidation processes of metallic Mn to Mn^{2+} that cause the Li_2O decomposition (**Figure 10b**). The rate capability of the working electrodes was performed at different current densities as shown in **Figure 10c**. The cell delivers average charge capacities of 377, 310, 273, 235, 180, and 104 mAh g^{-1}, at current rates of C/5, C/4, C/3, C/2, 1C, and 2C respectively with 1C = 1.87 A g^{-1}. When the current density was back to C/5, a charge capacity of 233 mAh g^{-1} was recovered. With a coulombic efficiency of 96, 98, 98, 99, 99, 98, and 99%, respectively.

4. Conclusions

In this chapter, we have reported the potential use of the transition metal oxides (TMOs) as promising anodes for LIBs. TMOs are characterized by their high theoretical capacities, suitable operation voltages, and a variety of chemical compositions. However, their practical use is still limited due to their poor electronic conductivity and high-volume expansion during the Li^+ insertion/extraction. To alleviate these challenges different approaches were suggested:

- The design of three-dimensional (3D) structures with hierarchical configurations,
- The elaboration of the TMOs at the nanometric scale,
- The production of hollow structures,
- The hybridization with carbonaceous materials besides the carbon coating approach.

Here, we have outlined how reducing the particle size to the nanoscale enhances the lithium storage reversibility for the conversion of metal oxides. $Mn_xCo_{3-x}O_4$ $0 \leq x \leq 3$, octahedral-like nanoparticles were prepared by a bio-inspired, cost-effective, and environmentally friendly synthesis method using metal nitrate as a precursor and alginate biopolymer as a sacrificial biotemplate. After a thermal treatment of the organic-inorganic compound at 600°C, the biopolymer degrades and the metal oxides with spinel structure are released. Electrochemical characterization of $Mn_xCo_{3-x}O_4$ $0 \leq x \leq 3$ demonstrated that these materials exhibit high capacities with good capacity retention and rate capability performances, making them potential anode materials for the next generation LIBs.

Acknowledgements

The authors wish to acknowledge Office Chérifien des Phosphates (OCP S.A.) and Mohammed VI Polytechnic University for financial support, and PCM2E laboratory of Tours University.

Author details

Loubna Hdidou[1], Fouad Ghamouss[1,2], Bouchaib Manoun[1,3], Hassan Hannache[1,4], Jones Alami[1] and Mouad Dahbi[1*]

1 Materials Science and Nano-engineering Department, Mohammed VI Polytechnic University, Ben Guerir, Morocco

2 Laboratoire PCM2E (EA 6296), UFR Sciences et Techniques, Université François Rabelais de Tours, Tours, France

3 Hassan First University Rayonnement-Matière et Instrumentation, Settat, Morocco

4 Laboratoire d'Ingénierie et Matériaux (LIMAT), Faculty of Sciences Ben M'sik, Hassan II University of Casablanca, Casablanca, Morocco

*Address all correspondence to: mouad.dahbi@um6p.ma

References

[1] Wang Q, Mao B, Stoliarov SI, Sun J. A review of lithium-ion battery failure mechanisms and fire prevention strategies. Progress in Energy and Combustion Science. 2019;**73**:95-131

[2] Stevens DA, Dahn JR. The mechanisms of lithium and sodium insertion in carbon materials. Journal of Electrochemical Society. 2001;**148**(8):A803. DOI: 10.1149/1.1379565

[3] Hassoun J, Bonaccorso F, Agostini M, Angelucci M, Betti MG, Cingolani R, et al. A lithium-ion battery based on a graphene nanoflakes ink anode and a lithium iron phosphate cathode. Nano Letters. 2014;**14**(8):4901-4906

[4] Zhang T, Han S, Guo W, Hou F, Liu J. Continuous carbon nanotube composite fibers for flexible aqueous lithium-ion batteries. Sustainable Materials and Technologies. 2019;**20**:e00096

[5] Yin H, Li Q, Cao M, Zhang W, Zhao H, Li C, et al. Nanosized-bismuth-embedded 1D carbon nanofibers as high-performance anodes for lithium-ion and sodium-ion batteries. Nano Research. 2017;**10**(6):2156-2167

[6] Xue C, Liu Y, Zhao J, Li X, Zhang J, Zhang J. Micro–mesoporous nitrogen-doped hollow carbon nanospheres as anodes for lithium-ion batteries with high-rate capability and outstanding cycling performance. Ceramics International. 2021;**48**:5434-5441

[7] Sun H, Del Rio Castillo AE, Monaco S, Capasso A, Ansaldo A, Prato M, et al. Binder-free graphene as an advanced anode for lithium batteries. Journal of Material Chemistry A. 2016;**4**(18):6886-6895. DOI: 10.1039/C5TA08553E

[8] Hansen S, Quiroga-González E, Carstensen J, Föll H. Size-dependent cyclic voltammetry study of silicon microwire anodes for lithium-ion batteries. Electrochimica Acta. 2016;**217**:283-291

[9] Kang K, Lee H, Han D, Kim G, Lee D, Lee G, et al. Maximum Li storage in Si nanowires for the high-capacity three-dimensional Li-ion battery. Applied Physical Letters. 2010;**96**(5):053110

[10] Wang L, Bao K, Lou Z, Liang G, Zhou Q. Chemical synthesis of germanium nanoparticles with uniform size as anode materials for lithium-ion batteries. Dalton Transactions. 2016;**45**(7):2814-2817. DOI: 10.1039/C5DT04749H

[11] Zhu J, Ding X. A facile one–pot synthesis of Sn/graphite/graphene nanocomposites as anode materials for lithium–ion batteries. Journal of Alloys and Compounds. 2019;**809**:151870

[12] Meng J-K, Fu L, Liu Y, Zheng G, Zheng X, Guan X, et al. Gas-liquid interfacial assembly and electrochemical properties of 3D highly dispersed α-Fe_2O_3@graphene aerogel composites with a hierarchical structure for applications in anodes of lithium ion batteries. Electrochimica Acta. 2017;**224**:40-48

[13] Zheng F, Yin Z, Xia H, Bai G, Zhang Y. Porous MnO@C nanocomposite derived from metal-organic frameworks as anode materials for long-life lithium-ion batteries. Chemical Engineering Journal. 2017;**327**:474-480

[14] Palmieri A, Spinner N, Zhao S, Mustain WE. Explaining the role and mechanism of carbon matrices in enhancing reaction reversibility of metal

oxide anodes for high performance Li ion batteries. Carbon. 2018;**130**:515-524

[15] Varapragasam SJP, Balasanthiran C, Gurung A, Qiao Q, Rioux RM, Hoefelmeyer JD. Kirkendall growth of hollow Mn_3O_4 nanoparticles upon galvanic reaction of MnO with Cu^{2+} and evaluation as anode for lithium-ion batteries. Journal of Physical Chemistry C. 2018;**121**(21):11089-11099

[16] Li ZH, Zhao TP, Zhan XY, Gao DS, Xiao QZ, Lei GT. High capacity three-dimensional ordered macroporous $CoFe_2O_4$ as anode material for lithium ion batteries. Electrochimica Acta. 2010;**55**(15):4594-4598

[17] Duan L, Wang Y, Wang L, Zhang F, Wang L. Mesoporous MFe_2O_4 (M=Mn, Co, and Ni) for anode materials of lithium-ion batteries: Synthesis and electrochemical properties. Materials Research Bulletin. 2015;**61**:195-200

[18] Cheng X, Li Y, Shi H, Lu J, Zhang Y. Rate-dependent electrochemical reaction mechanism of spinel metal oxide anode studied by in situ TEM. Journal of Alloys and Compounds. 2018;**763**:349-354

[19] Boebinger MG, Yeh D, Xu M, Miles BC, Wang B, Papakyriakou M, et al. Avoiding fracture in a conversion battery material through reaction with larger ions. Joule. 2018;**2**(9):1783-1799

[20] Cao K, Jin T, Yang L, Jiao L. Recent progress in conversion reaction metal oxide anodes for Li-ion batteries. Material Chemical Frontiers. 2017;**1**(11):2213-2242. DOI: 10.1039/C7QM00175D

[21] Lu Y, Yu L, David Lou XW. Nanostructured conversion-type anode materials for advanced lithium-ion batteries. Chemistry. 2018;**4**(5):972-996

[22] Park YJ, Lee KS, Shim J, Lee J-H, Kim Y, Son DI. Suppression of volume expansion by graphene encapsulated Co_3O_4 quantum dots for boosting lithium storage. Journal of Industrial and Engineering Chemistry. 2021;**95**:333-339

[23] Singh J, Lee S, Kim S, Singh SP, Kim J, Rai AK. Fabrication of 1D mesoporous NiO nano-rods as high capacity and long-life anode material for lithium ion batteries. Journal of Alloys and Compounds. 2021;**850**:156755

[24] Wang Z, Shi H, Yang S, Cai Z, Lu H, Jia L, et al. Oxygen vacancy-enriched Co_3O_4 as lithiophilic medium for ultra-stable anode of lithium metal batteries. Journal of Alloys and Compounds. 2021;**888**:161553

[25] Kong X, Zhu T, Cheng F, Zhu M, Cao X, Liang S, et al. Uniform $MnCo_2O_4$ porous dumbbells for lithium-ion batteries and oxygen evolution reactions. ACS Applied Material Interfaces. 2018;**10**(10):8730-8738

[26] Wu D, Zhang L, Zhang J, Zhang Z, Liang F, Jiang L, et al. Novel self-supporting multilevel-3D porous NiO nanowires with metal-organic gel coating via "like dissolves like" to trigger high-performance binder-free lithium-ion batteries. Microporous and Mesoporous Materials. 2021;**328**:111483

[27] Augustin M, Fenske D, Bardenhagen I, Westphal A, Knipper M, Plaggenborg T, et al. Manganese oxide phases and morphologies: A study on calcination temperature and atmospheric dependence. Beilstein Journal of Nanotechnology. 2015;**6**:47-59

[28] Li D, Yang D, Zhu X, Jing D, Xia Y, Ji Q, et al. Simple pyrolysis of cobalt alginate fibres into Co_3O_4 /C nano/microstructures for a high-performance lithium ion battery anode. Journal of Materials Chemistry A. 2014;**2**(44):18761-18766

[29] Koo B, Xiong H, Slater MD, Prakapenka VB, Balasubramanian M, Podsiadlo P, et al. Hollow iron oxide nanoparticles for application in lithium ion batteries. Nano Letters. 2012;**12**(5):2429-2435

[30] Jian G, Xu Y, Lai L-C, Wang C, Zachariah MR. Mn_3O_4 hollow spheres for lithium-ion batteries with high rate and capacity. Journal of Material Chemistry A. 2014;**2**(13):4627-4632. DOI: 10.1039/C4TA00207E

[31] Karunakaran G, Kundu M, Maduraiveeran G, Kolesnikov E, Gorshenkov MV, Balasingam S, et al. Hollow mesoporous heterostructures negative electrode comprised of $CoFe_2O_4@Fe_3O_4$ for next generation lithium-ion batteries. Microporous and Mesoporous Materials. 2018;**272**:1-7

[32] Xie Y, Qiu Y, Tian L, Liu T, Su X. Ultrafine hollow Fe_3O_4 anode material modified with reduced graphene oxides for high-power lithium-ion batteries. Journal of Alloys and Compounds. 2022;**894**:162384

[33] Wu L, Xiao Q, Li Z, Lei G, Zhang P, Wang L. $CoFe_2O_4$/C composite fibers as anode materials for lithium-ion batteries with stable and high electrochemical performance. Solid State Ionics. 2012;**215**:24-28

[34] Li S, Wang B, Liu J, Yu M. In situ one-step synthesis of $CoFe_2O_4$/graphene nanocomposites as high-performance anode for lithium-ion batteries. Electrochimica Acta. 2014;**129**:33-39

[35] Wang L, Zheng Y, Wang X, Chen S, Xu F, Zuo L, et al. Nitrogen-doped porous carbon/Co_3O_4 nanocomposites as anode materials for lithium-ion batteries. ACS Applied Materials Interfaces. 2014;**6**(10): 7117-7125

[36] Hdidou L, Khallouk K, Solhy A, Manoun B, Oukarroum A, Barakat A. Synthesis of CoFeO mixed oxides *via* an alginate gelation process as efficient heterogeneous catalysts for lignin depolymerization in water. Catalysis Science Technology. 2018;**8**(21):5445-5453. DOI: 10.1039/C8CY00576A

[37] Xing X, Liu R, Liu S, Xiao S, Xu Y, Wang C, et al. Surfactant-assisted hydrothermal synthesis of cobalt oxide/nitrogen-doped graphene framework for enhanced anodic performance in lithium-ion batteries. Electrochimica Acta. 2016;**194**:310-316

[38] Zhang L, He G, Lei S, Qi G, Jiu H, Wang J. Hierarchical hollow microflowers constructed from mesoporous single crystalline $CoMn_2O_4$ nanosheets for high performance anode of lithium ion battery. Journal of Power Sources. 2016;**326**:505-513

[39] Luo Y, Chen C, Chen L, Zhang M, Wang T. 3D reticular pomegranate-like $CoMn_2O_4$/C for ultrahigh rate lithium-ion storage with re-oxidation of manganese. Electrochimica Acta. 2017;**241**:244-251

[40] Wang M, Huang Y, Zhang N, Wang K, Chen X, Ding X. A facile synthesis of controlled Mn_3O_4 hollow polyhedron for high-performance lithium-ion battery anodes. Chemical Engineering Journal. 2018;**334**:2383-2391

[41] Pan X, Ma J, Yuan R, Yang X. Layered double hydroxides for preparing $CoMn_2O_4$ nanoparticles as anodes of lithium-ion batteries. Materials Chemistry and Physics. 2017;**194**:137-141

[42] Yan J, Zhang J, Su Y-C, Zhang X-G, Xia B-J. A novel perspective on the formation of the solid electrolyte interphase on the graphite electrode for lithium-ion batteries. Electrochimica Acta. 2010;**55**(5):1785-1794

Chapter 3

Perspective Chapter: Thermal Runaway in Lithium-Ion Batteries

Iñaki Lalinde, Alberto Berrueta, Juan José Valera, Joseba Arza, Pablo Sanchis and Alfredo Ursúa

Abstract

Lithium-ion batteries (LIBs) are becoming well established as a key component in the integration of renewable energies and in the development of electric vehicles. Nevertheless, they have a narrow safe operating area with regard to the voltage and temperature conditions at which these batteries can work. Outside this area, a series of chemical reactions take place that can lead to component degradation, reduced performance and even self-destruction. The phenomenon consisting of the sudden failure of an LIB, causing an abrupt temperature increase, is known as thermal runaway (TR) and is considered to be the most dangerous event that can occur in LIBs. Therefore, the safety of LIBs is one of the obstacles that this technology must overcome in order to continue to develop and become well established for uses in all types of applications. This chapter presents a detailed study of the general issues surrounding this phenomenon. The origin of the problem is identified, the causes are detailed as well as the phases prior to TR. An analysis is made of the most relevant factors influencing this phenomenon, and details are provided of detection, prevention and mitigation measures that could either prevent the TR or reduce the consequences.

Keywords: thermal runaway, safety, lithium-ion batteries, renewable energies, electric vehicles

1. Introduction

The fight against climate change, primarily due to greenhouse gas emissions, is now a reality. Two key strategies that are directed at tackling this phenomenon are an increase in the generation of electricity from renewable sources and vehicle electrification. Firstly, renewable energies (REs) represent an extremely effective solution, given that they are a clean and unlimited energy source that can substantially reduce the polluting emissions associated with power production. Secondly, electric vehicles (EVs) are the most effective instrument for cutting the emissions associated with transport, particularly in urban environments, and for improving the quality of life of mankind. Energy storage systems (ESSs) play a decisive role in the development of these renewable energy generation and electric mobility technologies. On the one hand, ESSs are essential for REs, permitting energy manageability and enhancing the integration of renewables into the power grid [1]. On the other hand, they play a key role in EVs, by giving these vehicles the run time required to make this form of transport a viable option.

Of the energy storage systems available on the market, lithium-ion batteries (LIBs) are the most widely used alternative in applications of this type. By the end of 2020, 71.1% of the operational energy storage project capacity in the world, not accounting pumped hydro storage, is based on LIBs [2]. LIBs offer attractive advantages, such as a high energy density (265 Wh/kg), high power density (340 W/kg), long lifetime (more than 5000 cycles), high efficiency (95%) and a low self-discharge rate (0.5–2.5% per month). However, their main drawback is their cost ($137/kWh) [3], which is still too high to be attractive in some applications. Despite the fact that substantial reductions have been achieved in recent years, further reductions are now being held back by the cost of the critical constituent materials such as cobalt, lithium and natural graphite [4]. Furthermore, given that these batteries have a limited safe operating area (SOA) with regard to the voltage and temperature conditions that they can withstand, a battery management system (BMS) is required. If the LIBs operate outside this safe window, certain undesirable chemical reactions may take place, possibly causing degradation, capacity loss or, in the worst case, the self-destruction of the cell.

The most dangerous phenomenon that can occur in a lithium-ion cell, and which may cause its self-destruction, is known as thermal runaway (TR). TR is a sharp increase in the temperature and pressure inside the cell, caused by a series of exothermic chemical chain reactions that feed one other, releasing heat and gases [5]. This phenomenon may be accompanied by the release of flammable and toxic gases and the risk of a fire or explosion. Moreover, one of the main characteristics of this phenomenon is that, despite the fact that it is initiated in one cell, it can then propagate to the rest of the cells making up the module and could even affect the whole battery [6].

To tackle TR, the cells incorporate safety devices. Furthermore, there is a BMS at a module or battery pack level, and on occasions, there is also a battery thermal management system (BTMS). On the one hand, the BMS is responsible for monitoring variables such as the voltage of each cell, the current through the battery and the temperature at some points of the battery. On the other hand, it monitors the battery state to check that it remains within its safe operating limits. By contrast, the BTMS controls the cooling or heating of the cell in order to ensure the right temperature, and it may be controlled by the BMS itself. Despite the safety devices installed at a cell and module level, the possibility of TR cannot be ruled out, in view of the accidents caused by this phenomenon [5, 7]. Furthermore, the number of accidents could increase over the next few years due to the boom in LIBs in applications such as REs or EVs [8]. Therefore, investigations directed at gaining a greater understanding of TR in order to take effective prevention, detection and mitigation measures are essential for the safe development of LIBs.

This chapter aims to make a general analysis of the key factors surrounding the phenomenon of TR in LIBs. It is therefore intended to serve as a guide for all interested parties, including the manufacturers, battery pack designers and even the users of LIBs. The different sections of this chapter explain in detail the TR mechanism, analysing its causes and its phases, during the actual phenomenon and also in the process prior to its initiation. In addition, the factors affecting the characteristics of TR are also indicated and the variables that may permit its early detection. Finally, this chapter looks at the resulting risks and hazards and possible prevention and mitigation strategies.

2. Thermal runaway in lithium-ion batteries

The characteristics of the TR occurring in a lithium cell depend on the type of cell in question. Specifically, the most relevant characteristic of the cell is its chemistry. The chemistry of a cell refers to the materials used in the internal components of the LIBs. The main components are the anode, cathode, electrolyte and separator. These materials provide the electrochemical properties of the cell and, to a larger extent, determine its level of safety as far as TR is concerned. A number of LIB chemistries are currently available on the market, with different characteristics with regard to rated voltage, energy density, lifespan, safety and cost. In general, those chemistries offering higher energy densities are the least safe [5]. However, as a result of the need to use LIBs in ever more demanding applications which require high energy densities, such as EVs, chemistries that are more unstable and unsafe are being used. For this reason, there is a strong emphasis on R&D directed at the development of batteries with the best electrical performance, yet without compromising their level of safety.

TR can be defined as a sharp increase in the cell temperature in a short period of time due to a successive series of exothermic chemical reactions that take place between the internal components of the cell. These unwanted reactions can be caused by abuse conditions such as overtemperatures or overvoltages, which break the stability of the internal components, or even by mechanical abuse, which can pierce the separator and produce an internal short circuit (ISC). The heat generated by these reactions is greater than the amount of heat that the cell is able to dissipate, thereby increasing the cell temperature. As the temperature rises, further chemical reactions occur between the internal components of the cell, causing irreversible damage. Although these reactions, occurring in the initial heating phases, degrade the cell, they still do not trigger TR. However, if the temperature continues to rise, then it finally reaches the TR onset temperature, causing some highly exothermic reactions which produce a tremendous release of energy in a very short space of time, leading to a sharp increase in the cell temperature and to its self-destruction. From that moment onwards, the phenomenon becomes uncontrollable, producing flammable and toxic gases which in turn increase the internal pressure of the cell. This overpressure causes the gases to be released through the pressure relief vents or due to a rupture of the casing, known as venting. When the inflammable gases released come into contact with the outside air, there is a risk of explosion or fire. Once TR has been initiated in a cell, the phenomenon may propagate to the other nearby cells forming part of the module and may even affect the whole battery if suitable containment measures are not taken. Propagation is one of the greatest risks derived from TR, leading it to spread to other healthy cells, exponentially increasing its hazardous potential.

Figure 1 shows the limits of what is known as the Safe Operating Area (SOA) of LIBs. The SOA is the voltage and temperature range in which a lithium cell can operate without causing undesirable reactions leading to accelerated degradation or ultimately to TR. As can be seen in the figure, higher temperatures can trigger a series of chain reactions leading to TR [9]. However, this phenomenon can also occur in other ways, given that charging at low temperatures or overcharging can also cause lithium plating. Lithium plating produces the growth of dendrites within the cell, even inside the safe voltage range if other conditions are satisfied, such as low temperatures and high charge rates [10, 11]. These dendrites could pierce the separator and cause an internal short circuit (ISC) that releases sufficient energy to trigger other exothermic reactions that lead to TR. In **Figure 1** it is also possible to observe how the reactions produce gases that increase the internal pressure, causing venting and the subsequent

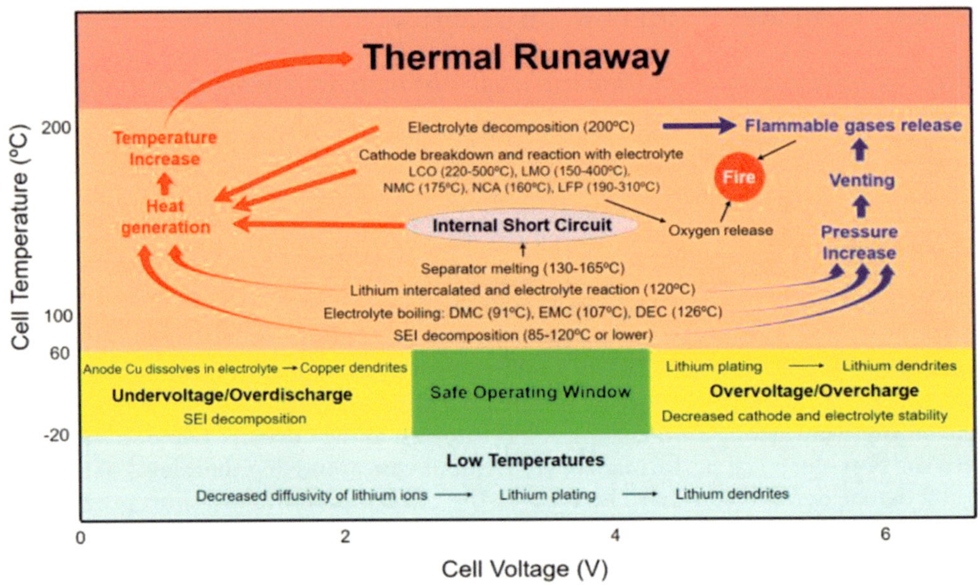

Figure 1.
Safe operating area (SOA) of LIBs and the phenomena that may occur outside of it.

release of these gases into the surroundings. It is therefore essential to maintain LIBs within the SOA to ensure that they remain safe.

In order to improve the safety of LIBs and mitigate the consequences of TR, a number of international commissions and organisations have drawn up and passed certain standards to evaluate their safety [12, 13]. These standards require LIBs to pass certain safety tests in different abuse conditions, including overcharge, over-discharge, overheating and mechanical impact, among others. The safety test standards are designed to try and ensure that certified LIBs have a sufficiently low risk of accident and, therefore, can be safely used in industrial or household devices. The battery safety test standards

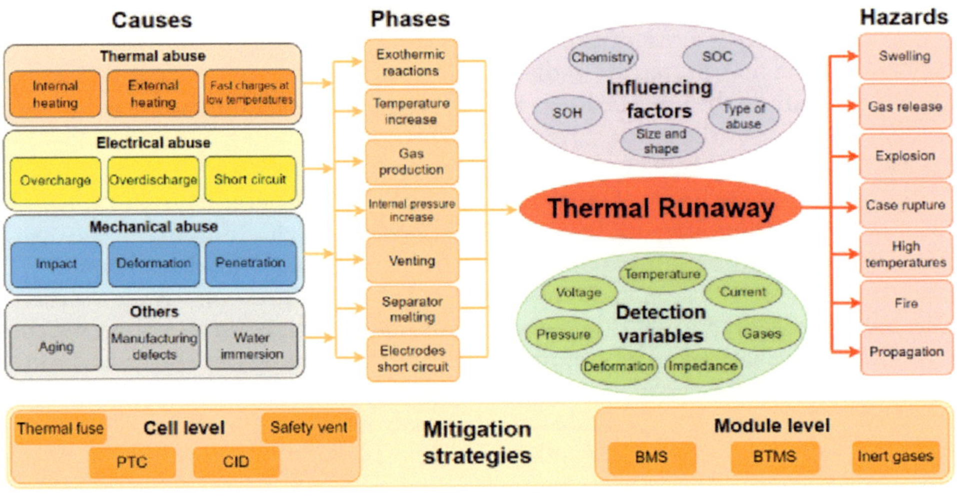

Figure 2.
General TR aspects in LIBs.

are constantly being updated and optimised, given that current tests cannot completely guarantee the safety of LIBs in practical applications. This is a serious problem, due to the fact that battery fires are reported to occur around the world. For this reason, it is essential to gain a complete understanding of the TR phenomenon in LIBs, thereby making it possible to improve the safety-related shortcomings of this technology.

The diagram in **Figure 2** shows the general TR aspects in LIBs. Each of these aspects is analysed in the following sections of this chapter. In the first place, this makes it possible to understand the origin of this phenomenon, based on its causes and the phases occurring. It is also possible to determine which factors influence the initiation of TR and their characteristics. It is also of vital importance to understand the resulting risks and hazards in order to ensure that suitable protective measures are established. Finally, it is crucial to analyse the variables that can help in the early detection of TR and have the prevention and mitigation measures in place, directed at preventing the initiation of TR and to reduce its risks and consequences.

3. Causes of thermal runaway

TR is triggered when the cell is operating outside its SOA. Certain chemical reactions are produced, depending on the voltage and temperature conditions. However, the fact that the cell is operating outside its SOA does not mean that TR is inevitable. This is because the relationship between heat generation and dissipation plays a fundamental role [7]. If the heat generated is greater than the rate at which it is dissipated, then this will lead to a subsequent increase in temperature, which will initiate further exothermic reactions, feeding the process, and which could ultimately trigger a TR. In contrast, if the heat dissipation is greater than the heat generated, then the cell temperature will drop, halting the process before TR is triggered [6].

During the normal operation of the battery, the cells generate heat due to inefficiencies in the charge and discharge processes. If the battery has been designed correctly, then this heat does not put the safety of the battery at risk. Therefore, with regard to the causes that could trigger TR, account is taken of the abuses that could expose the cells to operating conditions that are outside the established limits. It is therefore possible to distinguish three types of abuse: thermal, electrical and mechanical. It should also be pointed out that there are other possible causes that could lead to TR, such as cell degradation, impurities or defects during the manufacturing process or abuses such as immersion in water.

3.1 Thermal abuse

Temperature plays a key role in LIB performance and safety. This variable affects the material properties of the cells and, therefore, the processes and reactions taking place during charge and discharge. For this reason, it is crucial to control this variable, thereby making it possible to maintain the temperature of the cells within the safe operating range. A temperature that is either above or below this range could seriously affect LIB performance, and more seriously still, it may put its safety at risk.

3.1.1 Effect of low temperatures on LIBs

LIB performance is affected by low temperatures (<5°C in most cells). On the one hand, the electrolyte viscosity increases at low temperatures, reducing the mobility of

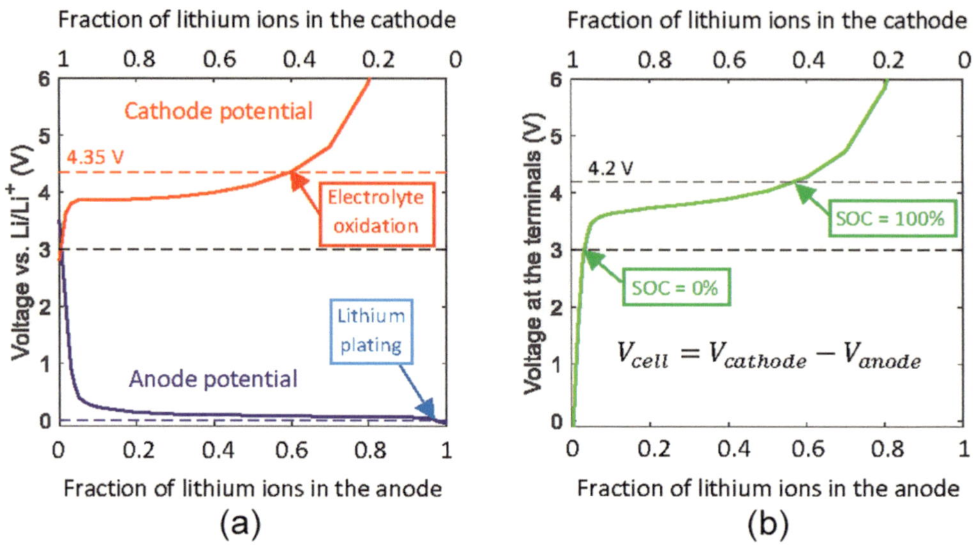

Figure 3.
(a) Voltage of each electrode vs. fraction of lithium ions in the anode; and (b) voltage at the cell terminals vs. fraction of lithium ions in the anode.

the lithium ions, thereby increasing the internal resistance of the cell [14]. On the other hand, the diffusivity of the lithium ions in the electrodes is reduced, restricting their intercalation in the crystalline structure. This causes a lithium concentration gradient in the electrodes, giving rise to a deviation in the intercalation voltage [15]. If the battery charge rate is greater than the anode diffusion rate, then the lithium intercalation voltage decreases even more. If this decrease reaches the voltage of Li/Li^+ (– 3.05 V vs. SHE, Standard Hydrogen Electrode), then the lithium ions are plated on the anode surface as metallic lithium [16]. **Figure 3a** shows at room temperature the equilibrium voltage in function of the fraction of lithium ions in each electrode [17]. The voltage at the cell terminals is the difference between these two voltages, the cathode voltage minus the anode voltage and is represented in **Figure 3b**. The horizontal axis shows the fraction of lithium ions in each electrode. As the fraction increases from 0 to 1, the voltage of the anode is reduced, whereas the cathode voltage is increased. When the cathode voltage goes over 4.35 V, then the electrolyte begins to oxidise. Furthermore, when the anode is full of lithium ions, which means that its voltage reaches zero, lithium plating occurs. The plated lithium reacts with the electrolyte through irreversible reactions, involving a loss of active material [18]. Therefore, on the one hand, this lithium plating causes the anode surface passivation, thereby reducing the maximum current that the cell can manage. On the other hand, the loss of active material leads to a reduction in the cell capacity [19]. In addition to these degradation phenomena, lithium plating causes the growth of dendrites, which can pierce the separator, giving rise to an ISC, which may trigger a TR. Low temperature charges therefore need to be performed with care, given that they can lead to the formation of metallic lithium dendrites that ultimately pierce the separator membrane.

3.1.2 Effect of high temperatures on LIBs

High temperatures can also put LIB safety at risk. In this case, the problem lies in the poor thermal stability of the materials which, once they reach certain temperatures, start to decompose or to react exothermically, releasing heat that accelerates the heating process and may trigger TR. Section 4 provides detailed comments on these reactions that take place in the cell as the temperature increases. The cause of this high temperature may be the external or internal overheating of the cell itself, when it exceeds its heat dissipation capacity. External heating can be transferred to the cell by conduction, convection or radiation and can either occur due to an excessive ambient temperature or when the cell is either in contact with or close to hot parts. On the other hand, internal heating refers to the heat generated by the cell during the charge and discharge processes or due to other chemical reactions.

One of the risks associated with overtemperature is that this situation may arise when the battery is not operating, and therefore, the TR detection systems may not be active. If a cell that is not operating is exposed to a high temperature that is above its safe operating limit, then this could ultimately trigger TR. For this reason, precautions need to be taken during the use of LIB and also during their transport and storage.

3.2 Electrical abuse

Just like temperature, voltage and current are crucial variables for the safety of LIBs. Depending on the cell chemistry, LIBs have maximum and minimum voltage values that determine their safe operating range. Outside this range, the materials are unstable, resulting in undesirable chemical reactions that degrade the cell and put its safety at risk. Given that there is a relationship between the cell voltage and its state of charge (SOC), to say that the cell has an overvoltage condition is to refer to the fact that an overcharge has occurred and, in the opposite case, an overdischarge [20].

As well as the voltage, a high current can affect the performance and safety of a LIB. For this purpose, manufacturers provide the maximum charge and discharge ratios that an LIB can withstand. As well as this type of overcurrent, which is sustained over time, the extreme case of an overcurrent is a short circuit that may either be caused outside (external short circuit) or inside the cell (internal short circuit).

3.2.1 Overcharge

Overcharge occurs when the battery is forcibly charged once the cut-off voltage has been reached. Overcharging causes a large amount of lithium ions to leave the cathode structure and to migrate to the anode. From the point of view of the cathode, the loss of a great many lithium ions may make it chemically unstable, and it may react exothermically with the electrolyte [21]. From the perspective of the anode, if the anode structure is unable to house all the lithium ions coming from the cathode, then these start to plate on the anode surface in the form of metallic lithium. The lithium plated may react with the electrolyte, causing an increase in the impedance and heat generated. In the worst-case scenario, it could lead to the formation of dendrites that could pierce the separator and cause an ISC. Finally, the commonly used electrolytes comprising $LiPF_6$ in an organic solvent mixture of EC-DEC-DMC have an electrochemical window that is limited to 4.5 V [22]. It is important to emphasise that a TR caused by an overcharge is more hazardous, due to the fact that the cell stores a greater amount of energy that can be released during the TR process. Moreover, as will be seen in Section 5.4, when the cell is overcharged, the TR trigger temperature drops, and it is therefore easier for the phenomenon to take place.

3.2.2 Overdischarge

Overdischarge means that the cell is discharged to under its lower voltage limit, which depends on the chemistry of the battery. Overdischarge presents less risks than overcharge, given that there is less energy stored in the battery, and therefore, if a TR is triggered, it will be less destructive. However, overdischarging can cause the decomposition of the Solid-Electrolyte Interphase (SEI), the rupture of the electrode materials and the formation of copper dendrites [23]. The SEI decomposition generates gases such as CO_2, methane (CH_4) and carbon monoxide (CO), causing the cell to swell [24]. When the battery is recharged, a new SEI is formed, consuming active lithium ions and electrolyte materials. The new SEI changes the electrochemical properties of the anode, leading to a capacity fade [25]. In the case of a deep overdischarge, once the potential of the anode reaches the copper oxidation potential (3.4–3.5 V vs. Li/Li^+), the copper current collector is oxidised [26]. The Cu^{2+} ions are solved in the electrolyte and may migrate through the separator and plate on the surface of the cathode [27]. When the cell is recharged, the solved copper is re-plated on the anode. If this process is repeated, metal dendrites can grow between the electrodes, possibly piercing the separator and creating an ISC [28], which could generate sufficient heat to trigger a TR.

3.2.3 Overcurrent

Based on the cell design, each manufacturer provides the maximum charge and discharge ratios that it is able to withstand without affecting its performance or safety. When there is no proper control of the current passing through the cell and the maximum ratios established are exceeded, then this produces a temperature increase due to greater heat generation [29]. This is due to the fact that the cell has been designed to dissipate a given amount of heat, and when an excessive current flows through the cell, the generation of heat is greater than its dissipation capacity. When the cell temperature increases above its safe range, this can trigger a number of reactions that degrade its materials and affect its performance. If the heat generation is maintained above the dissipation rate, then other reactions may be initiated that may ultimately lead to a TR.

3.2.4 External short circuit (ESC)

The ESC in an LIB takes place when there is a low impedance electrical connection between the cathode and anode current collectors on the battery. An ESC can occur for a number of reasons, such as a connection error, an insulation fault, mechanical abuse permitting the connection of the collectors and even by immersion in water or another substance that creates an electrical path. A high current peak occurs during an ESC, giving rise to great heat generation in a very short space of time and, therefore, a rapid increase in its temperature.

3.2.5 Internal short circuit (ISC)

The ISC is one of the most difficult phenomena to dissipate in LIBs given that, in contrast to the ESC, the electrical connection between two parts at a different potential takes place within the cell, making its prevention and detection more complicated. During the ISC, the heat generated by the Joule effect is inversely proportional to the ohmic resistance of the whole circuit. This resistance is a simplified way of representing the phenomena involved in heat generation in an equivalent circuit [30]. Its value largely depends on the active parts connecting the ISC. Structurally, the positive side of the cell includes a positive aluminium current collector with a coating of cathodic material, while the negative side consists of a negative copper current collector with its corresponding coating of anodic material. A separator membrane serves to electrically isolate the cathode from the anode. However, if the separator becomes damaged and is no longer able to maintain the isolation between the positive and negative sides, then this may cause an ISC.

Depending on the parts of the cell that come into contact, ISC can be classified into four types: Al-Cu, Ca-Cu, Al-An and Ca-An [31]. As far as the hazard level is concerned, these ISCs take the following order: Al-An > Al-Cu > > Ca-Cu > An-Ca [32]. This classification depends on the value of the equivalent short-circuit resistance which to a large extent determines the current produced and the amount of heat generated.

One of the reasons for which ISCs are exhaustively analysed in the TR study is that an ISC occurs during any TR situation. For example, when a cell is subjected to thermal abuse, the separator collapses when a given temperature is reached, leading to an ISC. Electrical abuse may lead to dendrite growth, piercing the separator and causing an ISC. Finally, mechanical abuse caused by deformation or penetration could rupture the separator and also lead to an ISC. Notwithstanding this, some studies have demonstrated that an ISC alone is not able to release sufficient energy to trigger TR [33]. The said work demonstrates that the ISC only releases a small amount of energy, given that the battery suffers a sharp increase in its internal resistance due to electrolyte consumption, reduction in electrolyte porosity and the separator shutdown. What is clear is that the separator prevents an ISC from occurring in the cell, and it may be damaged by a number of abuse conditions, as shown in **Figure 4**.

3.3 Mechanical abuse

When a lithium-ion cell is subjected to mechanical abuse conditions, such as impact, deformation or nail penetration, then this could pose a serious risk to safety. These factors are particularly relevant in applications such as EVs, which require

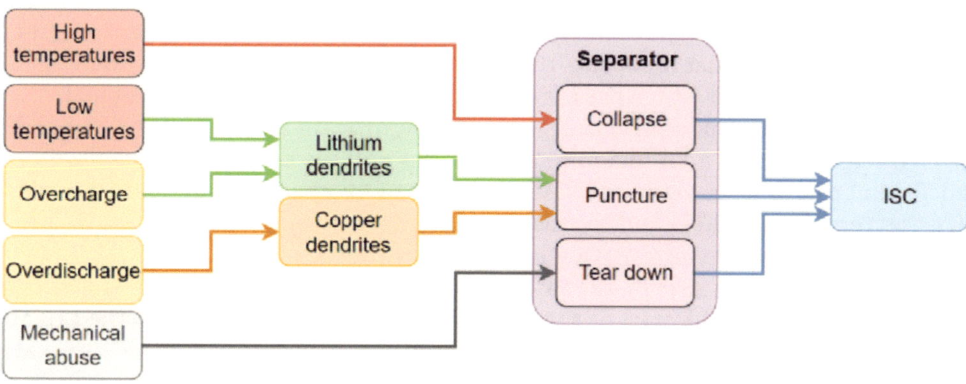

Figure 4.
Damage on the separator due to different causes.

an exhaustive study of the behaviour of LIBs in the event of any type of impact. Mechanical abuse may lead to cell deformation, affecting its internal structure and possibly leading to the rupture of the separator. In the event of a separator failure, a local ISC will occur, releasing heat due to the Joule effect. The heat generated increases the cell temperature and may even trigger exothermic reactions that increase the heat generation and produce gases. The large amount of heat generated in such a short space of time makes dissipation more complicated and may finally lead to the initiation of TR [34]. Therefore, the outer casing of the LIB plays a key role in preventing mechanical abuse and is responsible for withstanding mechanical loads and ensuring that there is no damage to the internal battery structure. In relation to TR, the chemistry of the components, the specific capacity of the battery and its state of charge play a decisive role [35]. A further determining factor is the type of mechanical abuse, given that the cell piercing may lead to the rupture of the casing, permitting the ingress of air which then reacts with the active components and the electrolyte, initiating strong exothermic oxidation reactions [35].

4. Phases of thermal runaway

Any TR phenomenon is associated with prior phases which, although they do not trigger TR, do affect the cell temperature, increasing the probability of the occurrence of those reactions that do cause TR. Each of the reactions between the internal cell components is initiated when the temperature reaches a certain value, while its duration is based on the rate at which the cell temperature increases [36]. Therefore, depending on the cause of the TR, these reactions may occur consecutively or simultaneously. For example, an LIB may be subject to mechanical abuse that causes an ISC, generating a large amount of heat in a very short space of time. In this case, the diverse reactions between the internal components will take place almost simultaneously. However, if the battery is exposed to external heating which slowly increases its temperature, then each reaction will take place when its starting temperature is reached.

In this section, in order to represent the temperature at which the chemical reactions are triggered, an external heating abuse of 2°C/min has been considered. **Figure 5a** shows the evolution of the cell temperature, its voltage and the gases released. These three variables are a good representation of the consequences of each of the TR stages. Furthermore, **Figure 5b** shows the temperature increase rate (dT/dt) in relation to the cell temperature. This is a widely used graph to represent the TR in LIBs, given that it permits the visual analysis of the exothermic behaviour and the moment at which the temperature that initiates the TR is reached. Furthermore, both graphs show the three most significant temperatures that permit the characterisation of a TR incident. These temperatures are identified as T_1, T_2 and T_3, and correspond to the onset temperature of exothermic reactions (T_1), the TR onset temperature (T_2) and the maximum temperature (T3). T_1, T_2 and T_3 provide extremely valuable information on the safety characteristics of a TR [37]. The greater the values of T_1 and T_2, the safer the cell, given that they indicate the safety margin available until the start of the exothermic reactions that degrade the LIB and the temperature of no return. On the other hand, the lower the value of T_3, the safer the cell, given that a lower value for this temperature indicates less energy released during the TR phenomenon which conditions, among other things, its potential propagation to the adjacent cells.

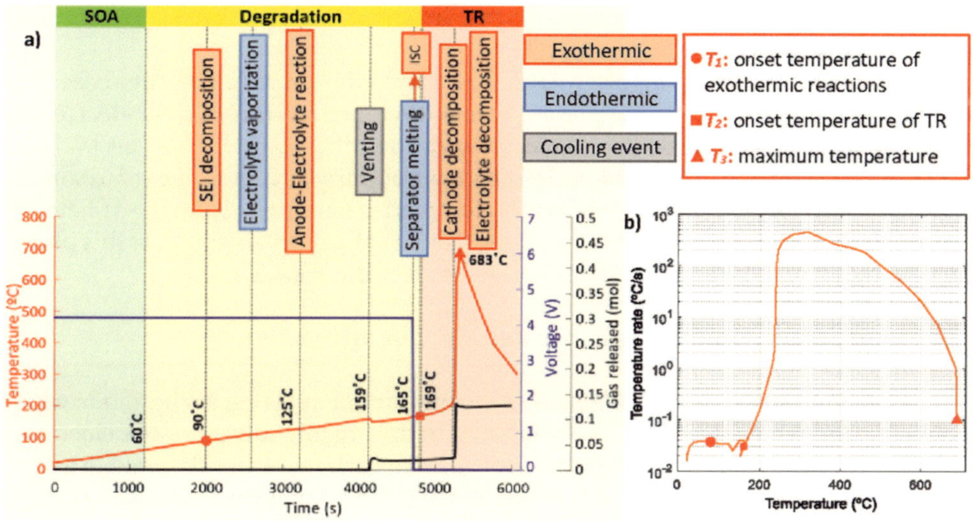

Figure 5.
Phases of TR. (a) Evolution of the temperature, voltage and gases release during an overheating test; and (b) temperature rate vs. temperature.

4.1 SEI layer decomposition

Once the cell temperature is outside the SOA, the decomposition of the SEI is the first exothermic reaction to take place. Therefore, the start of this reaction relates to temperature T_1, as can be seen in **Figure 5a**. Depending on the composition of the SEI layer, T_1 can be between 85 and 120°C [38]. However, it has also been confirmed in some cells that the decomposition of the SEI can start at lower temperatures, ranging from 60 to 80°C [28, 39]. Heat is released during the decomposition of the SEI, which increases the cell temperature. Moreover, flammable gases are produced such as ethylene (C_2H_4) and also some oxygen, due to the fact that the metastable components of the SEI such as $(CH_2OCO_2Li)_2$ are not thermally stable in the presence of lithium ions and may react exothermically [40]. In short, the decomposition of the SEI layer degrades the LIB and generates heat and gases, but does not in itself trigger the onset of TR. However, it does help to accelerate the temperature rise, which could lead to the start of new exothermic reactions. Furthermore, the loss of the protection of the SEI layer exposes the anode and means that it could react with the electrolyte.

4.2 Electrolyte vaporisation

The continued increase in the cell temperature leads to the vaporisation of the organic components of the electrolyte. The organic solvents making up the electrolyte, such as dimethyl carbonate (DMC), ethyl methyl carbonate (EMC) and diethyl carbonate (DEC), have boiling points of 91, 107 and 126°C, respectively [12]. On evaporation, these carbonates are converted into gases, which contribute to increase the cell pressure. Given that the volume remains constant, this causes the temperature to increase. In other words, despite the fact that the change of state from the liquid to vapour phase is endothermic, the energy absorbed is negligible in comparison to the increase in temperature caused by the pressure increase. In this phase, the temperature increases above the flashpoints of the gases produced by the electrolyte.

However, these gases do not combust, given that there is insufficient oxygen inside the cell to cause a fire [41].

Due to the utilisation of highly volatile and flammable organic-solvents-based electrolytes, new electrolyte systems are being developed to achieve safer electrolytes, such as ionic liquids, polymer-based electrolytes and solid-state electrolytes [42]. Ionic liquid electrolytes are characterised by low volatility and/or flame retardants properties. Polymer-based and solid-state electrolytes have the advantage of being free of liquid and organic solvents, avoiding the risk of cell leakage. Thus, this phase can be avoided or delayed with the use of other electrolyte system.

4.3 Lithium intercalated in the anode and electrolyte reaction

The decomposition of the SEI exposes the lithium intercalated in the anode which, at temperatures above 120°C, rapidly reacts with the organic solvent of the electrolyte [38]. This reaction leads to the formation of lithium carbonate and flammable hydrocarbons (ethylene, ethane and propylene). During the temperature rise, a second SEI layer may be formed, which again decomposes as a result of the above-mentioned reactions, with an exothermic peak at a temperature of 230°C [43].

4.4 Cell venting

The formation of gases in the cell interior causes the internal pressure to increase. Normally, the cells are fitted with devices that relieve the pressure when high values are reached, as this would involve a risk of explosion. When the safety vent is activated, the gases produced are released into the surroundings and the internal pressure is reduced. If the cell does not have a built-in safety vent, then the venting may occur through the rupture of the casing at its weakest point, as is the case with the pouch cells. As can be seen in **Figure 5a**, this process lowers the cell temperature due to the Joule-Thomson effect. This drop in temperature is due to the release of the hot gases, which reduces the pressure while maintaining the volume constant and therefore the temperature drops. This event takes place before the TR and, despite the drop in temperature, account should be taken of the fact that it involves a release of inflammable gases which, in contact with the oxygen in the air, could combust, causing an increase in the ambient temperature. This could accelerate the heating of the adjacent cells, favouring the propagation of TR. Finally, it is important to emphasise the fact that the amount of gases released during venting is significantly lower than those released in the TR itself, as can be seen in **Figure 5a**.

4.5 Separator melting and ISC

The separator consists of a porous membrane intended for two principal purposes: to allow ions to pass through and to electrically isolate the electrodes. Therefore, the separator material has high ionic conductivity and low electrical conductivity. At present, polyethylene (PE) and polypropylene (PP) are the most widely used materials for the separator. PE melts at 130°C, while PP does so at 165°C. Once melted, the separator membrane can no longer perform its function, triggering an ISC. As can be seen in **Figure 5a**, at the time when the separator melts, causing the subsequent ISC, the cell voltage abruptly drops to zero. The separator melting process is endothermic and absorbs energy during the change of state. However, the energy absorbed is negligible in comparison with that produced by the ISC.

4.6 Rupture of the cathode material and reaction with electrolyte

At high temperatures (175–250°C), the active material of the cathode (generally an oxide) decomposes exothermically, releasing oxygen. Both the oxygen released and the actual cathode material react with the electrolyte at this temperature. This chemical reaction is highly exothermic, causing the sharpest temperature increase in the TR process [38]. This reaction and the reaction between the intercalated lithium in the anode and electrolyte and the ISC caused by the collapse of the separator are the main triggers of the TR phenomenon in LIBs [6]. Therefore, the value of T_2 will be influenced by the weakest among the cathode, anode and separator.

4.7 Electrolyte decomposition

Both the anode and the cathode materials react with the electrolyte during TR. When the cell temperature exceeds 200°C, the electrolyte decomposes exothermically [38]. However, account should be taken of the fact that certain organic solvents in the electrolyte evaporate at temperatures below 100°C, thus, at 200°C, part of the electrolyte has already evaporated. The electrolyte decomposition produces flammable and toxic gases such as HF from the decomposition of $LiPF_6$ salt [44]. The generation of CO_2, C_2H_4 and HF, among other gases, has been documented in the decomposition of electrolytes comprising $LiPF_6$ and organic carbonates [45].

4.8 Reactions at higher temperatures

Other exothermic reactions that take place at very high temperatures have also been documented. At 220°C, the lithium intercalated in the negative electrode starts to react exothermically with the fluorine-based binder, usually polyvinylidene fluoride (PVDF) [38]. At temperatures above 300°C, metallic lithium can form from the decomposition of the stable components of the SEI layer (LiF and Li_2CO_3), and the graphite portion of the anode can exothermically decompose [43]. However, these take place following initiation of TR, and they have less influence in comparison with the aforementioned reactions [5].

5. Factors influencing thermal runaway

5.1 Chemistry

The chemistry of the batteries determines the characteristics of their TR, given that the reactions involved depend to a large extent on the materials used in the internal components. Depending on the type of chemistry, it is possible to find cells with a higher or lower degree of thermal stability, and therefore, they are more or less prone to TR. As discussed above, there are currently a number of different types of LIB chemistries on the market, which differ with regard to their electrochemical performance, lifespan and safety. In general, the name is determined by the active material used in the cathode. However, even when the same material is used for the cathode, the cells made by different manufacturers may behave differently with regard to the TR phenomenon, depending on the materials used in the rest of the internal components.

The active material of the cathode is either a lithium compound (LCO, LMO, NMC, NCA, LFP...) or a combination of these. The thermal stability of this material is crucial to determine the safety of the cell. Thermal stability refers to the temperature at which the cathode material decomposes and releases the oxygen contained in its structure. When this oxygen is released, it reacts with the electrolyte, generating heat and even igniting [46]. Cells with an LFP cathode are the safest, given that the strong phosphate bonds prevent the release of oxygen. By contrast, transition-metal oxides have weaker bonds, causing oxygen to be released at lower temperatures.

With regard to the anodic material, graphite is the most widely used due to its good performance and low cost [47]. However, the formation of an SEI layer is required, to prevent it from reacting with the electrolyte. As has been seen, the SEI layer decomposes at temperatures close to 90°C and is the first exothermic reaction to release heat and gases, which also gives rise to the reaction between the electrolyte and the lithium intercalated in the anode. Additionally, the SEI is an element that, to a large extent, suffers the effects derived from ageing. For this reason, other materials are available for the anode, with lithium titanate (LTO) as the most popular commercial alternative. An LTO anode does not need an SEI layer, making it safer. However, the intercalation voltage of lithium in the LTO is higher than that of graphite, leading to cells with a lower voltage and, therefore, with a lower energy density.

Finally, the electrolyte is generally composed of an organic solvent, a lithium salt and other additives. The materials used and their combination determine the stability of the electrolyte. Given that most of the electrolyte components are flammable, manufacturers generally add certain additives such as flame retardants for enhanced safety.

Table 1 provides a summary of the main characteristics of the lithium batteries, based on their chemistry. The last row indicates their safety characteristics. As can be observed, the LTO batteries are the safest, followed by the LFP ones. The LCO batteries are the ones with the lowest safety level. With regard to energy density, an inverse relationship with safety can be seen, whereby the safest batteries are also those with the lowest energy density and vice versa.

Abbreviation	LCO	LMO	NMC	LFP	LTO
Full name	Lithium cobalt oxide	Lithium manganese spinel	Lithium nickel cobalt manganese oxide	Lithium iron phosphate	Lithium titanate-oxide
Cathode	$LiCoO_2$	$LiMn_2O_4$	$Li(Ni_xMn_yCo_z)O_2$	$LiFePO_4$	LMO, NCM, LFP
Anode	Graphite	Graphite	Graphite	Graphite	$Li_4Ti_5O_{12}$
Cell voltage (V)	3.7–3.9	4.0	3.8–4.0	3.3	1.8–2.5
Energy density (Wh/kg)	150–190	100–160	140–200	90–140	70–85
Cycle life (cycles)	500–1000	1000–1500	1000–2000	≥ 2500	6000–27,000
Safety ranking	5	3	4	2	1

Data obtained from [9].

Table 1.
Main characteristics of various types of LIBs.

5.2 Size and shape

The cell size directly affects the energy released during the TR. A study conducted by Son et al. [48] analyses the influence of this factor for three LCO-graphite pouch cells with capacities of 33, 1000 and 3300 mAh. In the heating ramp test conducted in this publication, TR occurred in the 1000 and 3300 mAh cells, while this phenomenon did not occur in the cell with a lower capacity. The authors consider that this is due to the greater proportion of inactive materials in relation to the active materials in the lower capacity cell. Furthermore, they observed that the TR triggered in the 1000 and 3300 mAh cells had different characteristics, given that the cell with the greater capacity released a higher amount of energy, causing TR to be triggered earlier.

In relation to its shape, lithium-ion cells can be divided into three types: cylindrical, prismatic and pouch. Each shape has certain pros and cons with regard to its manufacturing process, assembly in the battery pack and storage capacity, built-in safety devices and safety from TR. With regard to TR, it is important to mention that the cylindrical and prismatic cells have a metal casing that allows them to withstand greater internal pressure before bursting than the pouch cells, whose plastic-aluminium laminate casing easily deforms when gases are generated inside the cell. The studies published report that pouch cells are able to withstand pressure values ranging between 223 and 411 kPa, while cells with a metal casing withstand pressures ranging from 386 to 855 kPa [49]. This causes venting to take place earlier in the pouch cells, allowing the gases to be released into the surroundings.

5.3 Type of abuse

TR is also affected by the type of abuse triggering it. As discussed in Section 3, there are different situations which give rise to TR. Therefore, depending on the type of abuse, the behaviour of the cell is different, and this determines the characteristics of the TR. This means that parameters such as the temperature at which TR is triggered, its maximum temperature, duration, venting and possible fire, may all vary depending on the underlying cause. This fact adds greater complexity to the characterisation of this phenomenon and makes its more unpredictable. For this reason, the correct experimental characterisation of TR involves the performance of different tests, including various possible causes. **Figure 6** shows the main results of an experimental study to analyse the evolution of the temperature and voltage of an NMC pouch cell with regard to three types of abuse: overtemperature (2°C/min), overcharge (1C) and nail penetration [49]. Firstly, looking at the horizontal axis of the figures, it can be observed that each test has a different duration. This shows the length of time during which the cell is able to withstand each abuse condition before the onset of TR. Thus, in the nail penetration test, TR is shown to occur just a few seconds after the cell is pierced. On the other hand, in the overcharge test, TR is triggered at around 1600 seconds and, for overtemperature, at 5000 seconds. Analysing the cell temperature for each test, the maximum temperature reached is observed to vary between 823°C for the overtemperature test and 760°C for the overcharge test. However, even more significant is the difference between the temperature at which TR is triggered in each test, being 206°C for the overtemperature test, around 100°C for the overcharge and just 30°C for the nail penetration. This clearly demonstrates the influence of the type of abuse on the characteristics of the TR phenomenon in LIBs, making its detection and prevention more complex.

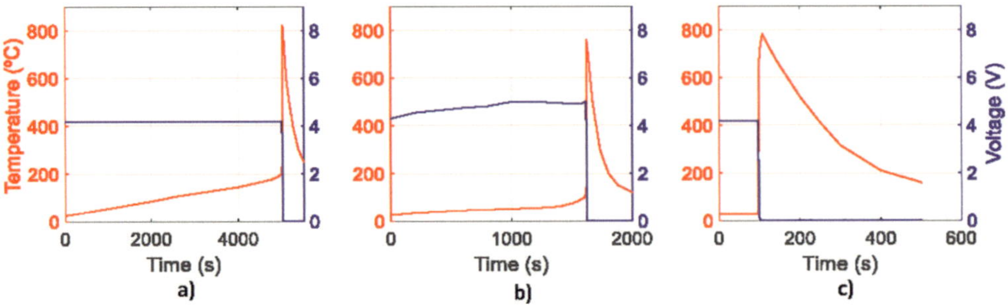

Figure 6.
Temperature and voltage evolution of the cell under different type of abuse. (a) Overheating; (b) overcharge; and (c) nail penetration. Data obtained from [49].

5.4 State of charge (SOC)

The battery state of charge (SOC) is also a determining factor in the analysis of TR. Firstly, the higher the SOC, the higher the amount of energy stored in the cell and, therefore, the greater the severity of TR. This is due to the release of a greater amount of energy during the phenomenon, resulting in a higher temperature T_3. Furthermore, the SOC also affects the value of the three characteristic temperatures that define the TR [50]. A higher SOC means an anode with a higher lithium content—lithiated anode—and a cathode with a lower lithium content—delithiated cathode. On the one hand, the lithiated anode has a reduced stability, which favours the reaction between the intercalated lithium and the electrolyte. This favoured exothermal reaction is triggered at lower temperatures for higher SOC, thereby reducing the onset temperature of exothermic reactions, T_1. On the other hand, the evolution of the TR is mainly related to the reactivity of the cathode with the electrolyte, given that the decomposition of the cathode material releases oxygen that reacts with the electrolyte. Given that the layered transition-metal oxide becomes unstable in the delithiated state—higher SOC, the onset temperature of TR, T_2, is also decreased. In addition, the TR process releases more heat with higher SOC, thereby increasing the maximum temperature during the event, T_3. In this way, it is important to emphasise that, if the cell is overcharged, it may have lower trigger temperatures [51]. In short, the higher the cell charge, the more prone it will be to TR and the worse the consequences. For this reason, during the transport and storage of LIBs, it is recommended to keep them at low SOC levels in order to reduce the risk of TR. This can be seen, for example, in standard UN 3480 which establishes a maximum SOC of 30% for the air transport of LIBs.

Figure 7 provides a compilation of the overtemperature test results conducted on different types of cells at different SOC, published in the literature [51–54]. **Figure 7a** shows the relationship between temperature at the onset of TR (T_2) and the SOC, observing a decreasing trend as the SOC increases. In contrast, **Figure 7b** shows the relationship between the maximum temperature (T_3) and the SOC, observing an increasing trend, as described in the paragraph above.

5.5 State of health (SOH)

Certain irreversible reactions take place throughout the useful life of an LIB, leading to cell degradation and loss of performance. The degradation primarily affects the

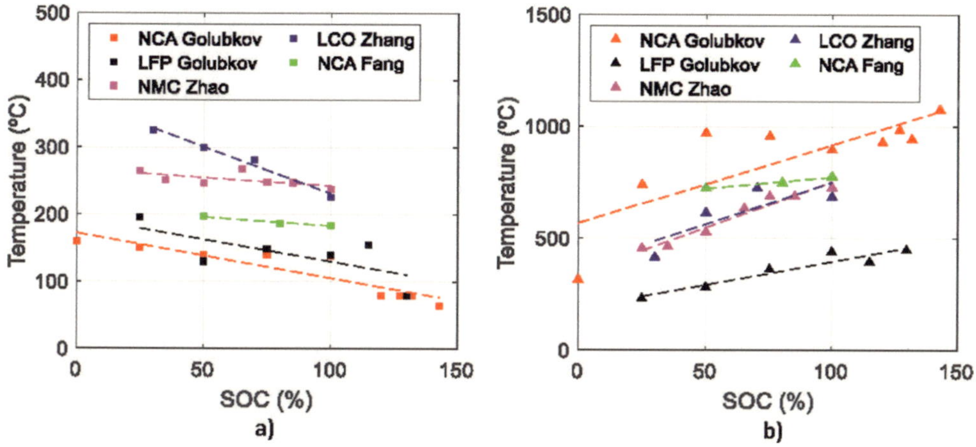

Figure 7.
SOC influence on TR (a) T2 vs. SOC; and (b) T3 vs. SOC. Data obtained from [51–54].

anode and, specifically, the SEI layer. The analysis of SOH as an influencing factor in TR is currently taking on particular importance given the growing interest in second life applications for lithium-ion batteries [55–57]. In this second life, due to the fact that the SOH is below the design value, this could give rise to phenomena for which the cell has not been designed.

It has been confirmed that, when LIBs age due to high temperature cycling, there is an improvement in their thermal stability, with an increase in T_1 and a reduction in heat generation [50]. This is due to the fact that the high temperature permits the formation of a more stable SEI layer. On the other hand, ageing by low-temperature cycling has the reverse effect, with a decrease in T_1 and T_2 and an increase in the self-heating rate.

Lithium plating may also occur in the form of dendrites which, as described in Section 3.2.1, may pierce the separator and cause an ISC that triggers TR. For this reason, an aged cell with a lower SOH is less safe with regard to the initiation of TR. However, as the SOH decreases, so does the amount of active materials available in the cell, thereby reducing the amount of heat released in a possible TR [58].

6. Hazards of thermal runaway

TR is the most dangerous phenomenon that can occur in an LIB, causing its self-destruction. However, the hazards derived from TR are not solely limited to the inoperability of a cell. As a result of a TR, the LIBs are exposed to extremely high temperatures which could cause burn injuries to persons or affect nearby equipment. The formation of gases in the cell interior increases the pressure, which could lead to venting, the rupture of the cell casing or an explosion. In addition to flammable gases, highly toxic gases are also generated such as hydrogen fluoride (HF) and phosphoryl fluoride (POF_3) [59]. The combustion of the released gases is dependent on meeting the three conditions making up what is known as the fire triangle. The fire triangle is the analogy represented in **Figure 8a** and which makes it possible to understand the three components required for combustion to take place: a fuel, an oxidising agent such as oxygen and energy or heat. All three elements are present in LIBs, and

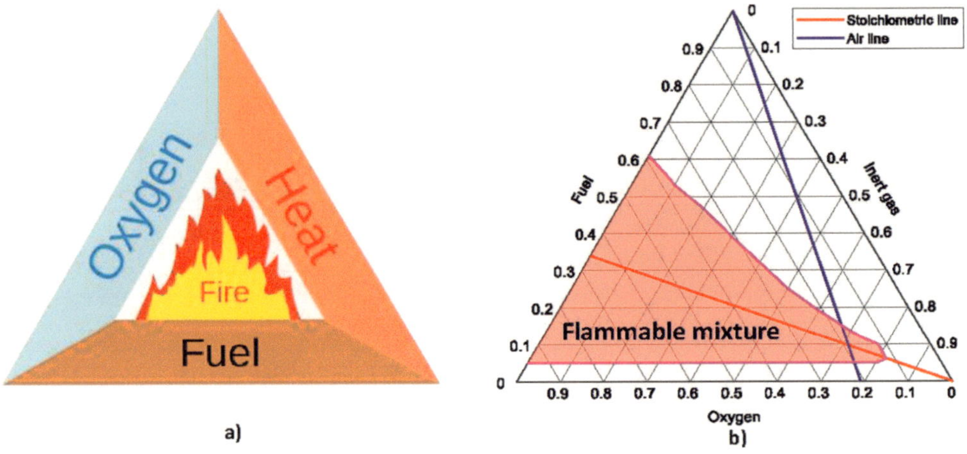

Figure 8.
(a) Fire triangle; (b) flammability diagram.

therefore, there is a probability of fire. The inflammable gases produced are the fuel. Oxygen is present either in the outside air or in the active material of the cathode, which is generally an oxide. Finally, heat is either generated by the actual exothermic reactions taking place, or it could also come from an ignition source such as a spark that may be produced as a result of the friction between the high-speed release of gases and the safety vent. It could also be due to an arc generated by the shorting of the electrodes during the TR [5]. Despite the fact that the three conditions of the triangle are met, combustion is not assured, given that the three elements must be combined in the right proportions for a fire to occur. Whenever any of these conditions are not met, then there can be no combustion. This can be extremely important with a view to extinguishing a fire caused by TR in a LIB [60]. In **Figure 8b** a ternary plot represents the flammability of mixtures of methane (fuel), oxygen and nitrogen (inert gas). Only when the concentrations of all three substances are in the red zone combustion can occur. The blue line is the air line representing all the possible mixtures of the substances under air conditions, from 79% of nitrogen and 21% of oxygen to 100% of methane. The red line is the stoichiometric line, and as the mixture is closer to that line, the combustion is more likely.

Another major thermal runaway hazard is its ability to propagate to other nearby cells, causing faults in "healthy" cells that probably would not have been affected by the abuse causing the TR in the source cell. The propagation of TR significantly increases the risk to the installation and to persons, given that the consequences of TR are multiplied by each cell affected. Propagation occurs through the transmission of heat from the source cell, either through conduction, convection or radiation, raising the temperature of the adjoining cells and initiating the exothermic reactions that lead to TR. The cell itself, as well as the gases released and the materials ejected, is responsible for the said transmission of heat. If the appropriate measures are not taken, once TR has been initiated in a cell, then it may be impossible to prevent propagation. Therefore, it is essential to take the phenomenon of TR into account in the battery design, not only at a cell level but also at a battery pack level, in order to have the necessary measures in place to reduce or prevent propagation.

7. Methods for the early detection of TR

In view of the hazardous nature of TR, it is essential to have devices in place that permit the detection of TR before it can even begin. New regulations such as GB 38031-2020 require the installation of a system in EVs to alert passengers of a battery fault at least 5 minutes in advance. In order to achieve this objective, it is necessary to have an in-depth knowledge of the behaviour of certain measurable variables that could provide information on the potential onset of TR. At present, the key LIB variables measured by the BMS are voltage, current and temperature. The monitoring of these variables permits adequate battery management, maintaining its operating point within the SOA. However, situations may arise in which both the BMS and the sensors may fail and may not be useful in guaranteeing the safety of the LIBs and the early detection of TR. With this in mind, other variables are now being investigated and which are not generally measured by current BMSs, but which are of great interest for the purpose of TR detection. The most important variables are discussed below, indicating their pros and cons.

7.1 Temperature

It has already been mentioned that the LIB temperature is a fundamental variable to be considered. The possibility of triggering TR can to a large extent be reduced by monitoring the temperature to ensure that it remains within the established safety range. Not withstanding this, situations may arise in which temperature is not a reliable variable, such as TR resulting from mechanical abuse or due to dendrite growth. If no account is taken of faults of this type, then temperature can be the most reliable variable for the early detection of the onset of TR. Once the cell temperature has exceeded the recommended limits, an alert can be given regarding the malfunctioning of the battery with sufficient time to prevent the temperature from continuing to rise, thereby avoiding exothermic reactions, cell degradation and the generation of more heat. This would therefore interrupt the path leading to chain reactions that ultimately trigger TR.

Fortunately, BMSs generally incorporate a battery temperature measurement function. Nevertheless, most BMSs do not monitor the temperature of all the battery cells. Instead, the temperature sensors are generally placed in strategical positions in order to reduce their number. In addition, significant thermal gradients may develop inside LIBs during charge and discharge, which are more pronounced with increased cell size and current rate [61]. As a result, it may happen that the temperature of a cell goes outside the safe range yet is not detected by the sensors until it is too late.

Moreover, a further drawback is that the sensors measure the surface temperature of the batteries, while the internal temperature may be greater and the surface-mounted sensor may take some time to detect rapid variations in this variable. **Figure 9** shows the difference between the internal temperature and that measured on the cell surface during rapid charge or discharge [62]. It can be observed that the surface temperature remains in the range between 20 and 40°C, while the internal temperature can exceed 60°C. The internal temperature is the most relevant, given that this is where the chemical reactions between the cell components take place. For this reason, some studies publish methods to measure the internal temperature by means of specific sensors [63]. The problem with these internal sensors is that they may affect the performance of the cell itself. What is more, when there are a

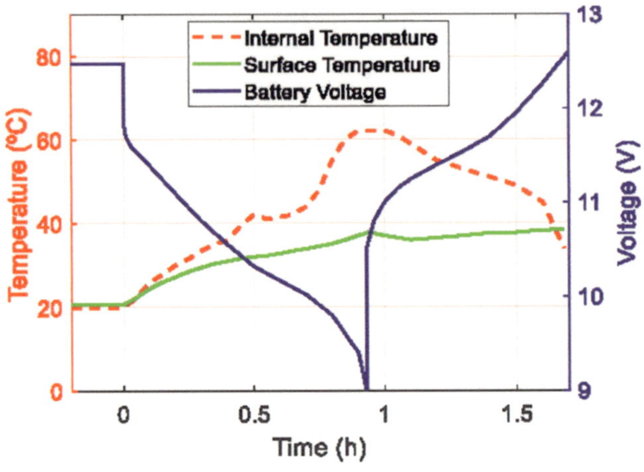

Figure 9.
Internal and superficial temperature during fast charging and discharging. Data obtained from [57].

very large number of cells, it can be expensive to place temperature sensors in each one. Currently, the techniques of internal temperature monitoring can be divided in two main themes of hard sensors and soft sensors [64]. In this reference, the hard sensors include sensors that need to be inserted into the cell, such as thermocouples, thermistors, resistance temperature detectors and optical fibre sensors or contactless methods such as EIS, which will be detailed in Section 7.7. On the other hand, soft sensors are estimators or observers that can estimate the internal temperature from superficial measurements using specific models.

7.2 Voltage

Just like temperature, voltage is another fundamental variable for LIBs, and manufacturers establish a safety range. Therefore, by monitoring the voltage, cells can be maintained at a safe operating point. The possibility of TR caused by overcharge or overdischarge can be avoided by monitoring the voltage (provided that the sensors and BMS are working correctly). However, in view of the results published in the literature for other types of abuse, such as external heating, voltage may not be indicative of the presence of an abuse situation that could trigger a thermal runaway. This is due to the fact that the voltage may not be affected by this type of abuse and a sharp voltage drop may only be detected once the separator melts and the subsequent ISC. However, this event generally occurs at temperatures of between 135 and 165°C, which could be a very late detection of the abuse. **Figure 10** shows the results of three identical tests on the same type of cell, where it can be seen how the voltage behaviour varies in each test [33]. The results of the tests shown in **Figure 10b** and **c** show how the voltage would have provided an early detection of TR with more than 300 seconds to spare. However, in the case of **Figure 10a**, the voltage drop is detected practically at the same time as the start of the TR, and there was therefore no advance warning.

7.3 Current

In the same way as voltage and temperature, this is a variable that is already measured by present-day BMSs, and so there is no need to add new hardware. Given

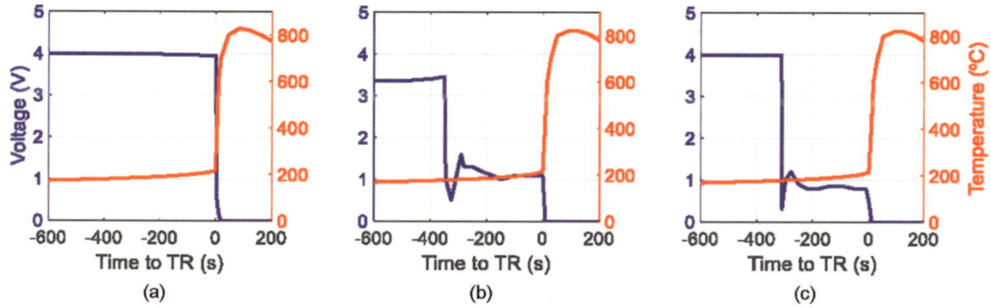

Figure 10.
Temperature and voltage evolution of the cell in three identical overtemperature tests. Data obtained from [28].

that its measurement makes it possible to monitor the battery charge and discharge, it is useful in normal operation in order to avoid excessive currents that may affect the internal processes taking place in the LIB. It also allows for the detection of external short circuits involving large current peaks. However, the main drawback is that the current is generally measured for the entire battery pack. This is because, when the cells are connected in series, the same current passes through all the cells. This may lead to a situation in which, although there is a fault in one cell and an ISC occurs, if the rest of the cells connected in series are operating correctly, then the system will not detect this. It is therefore difficult to detect a fault in large battery systems through current measurement. All the same, there are methods to detect an ISC in a cell in a battery system with cells connected in series and parallel by using $2{\cdot}(N_s - 1)$ ammeters, where N_s is the number of cells connected in series [65].

7.4 Pressure

The gases generated in the initial phases of TR, caused by the decomposition of the SEI and the vaporisation of the organic solvents in the electrolyte, create an increase in pressure in the cell interior. This, together with the temperature increase, creates a considerable increase in the internal pressure. This is therefore a variable that provides extremely reliable information on the formation of gases as a result of some type of abuse to the cell, thereby providing very early detection of TR, given that it would be detected at the start of the exothermic reactions that produce gases. Notwithstanding this, it is not easy to measure the internal pressure of the cell without modifying its structure or affecting its performance. On the other hand, it may be more feasible to contain the cells in sealed covers, such as an external casing that permits the mounting of a pressure sensor to detect the release of gases by the cell under abuse. Unfortunately, in this case, gases would be detected when they are released, rather than when they are produced, which would delay the detection of TR.

7.5 Deformation

Although this detection method is related to the internal pressure measurement, in this case the measurement would be taken from the cell exterior. A force or deformation sensor may be able to detect the increase in volume of the cells caused by an increase in internal pressure. This technique offers better results in pouch cells than in cylindrical or prismatic ones, given that they are subject to greater changes in volume during normal operation. The presence of gases produces a larger swelling in pouch cells, so that the measurement of this deformation provides detection prior to venting, and it may be possible to prevent the release of toxic and flammable gases into the surroundings. However, the drawback is the need for individual sensors for each cell, particularly in those applications in which there are hundreds or thousands of cells, which could make the battery more expensive.

7.6 Gas release

The chemical reactions that take place between the internal components of the cells result in the production of gases, thereby increasing the internal pressure until the gases are finally released into the surroundings. Given that venting is an event that takes place prior to TR, with the installation of gas sensors that are capable of detecting the presence of released gases, then early detection is possible in order to avoid TR [66]. The numerous articles in specialised literature on TR have analysed the composition of the gases released during venting and those released in TR. During TR, a greater amount of gases is released in comparison with venting [67]. In most publications in the literature, the composition of the gases is generally measured once thermal runaway has occurred. As this can vary considerably with the composition of the gases released during venting, with a view to the early detection of this phenomenon, it is essential to analyse the venting process. In general, the gases with the highest concentrations include CO_2, CO, H_2 and hydrocarbons from the vaporisation of the electrolyte organic solvents (DMC, DEC, EC) [49, 68, 69]. In general, the gases produced during venting are dominated by the used electrolyte mixture, whereas while TR, the side reactions are more significant [6, 70].

The main drawback to this detection method is the need for venting to occur sufficiently in advance of the onset of TR. Furthermore, the time between the venting and TR generally varies and depends on a number of factors, such as the abuse conditions, SOC and type of cell [49, 51]. Possibly, at the time when venting is detected, it may be too late to prevent the initiation of TR. Moreover, there is also the possibility that there is no prior venting before TR is triggered in the cell, such as when TR is caused by mechanical abuse, when it can be triggered in just a few seconds [49].

Unfortunately, venting can release flammable gases (which can combust on contact with the air) and toxic gases (which can be harmful to human health). Therefore, account should be taken of the consequences of the detection method failing to detect the venting in time, and that the cell may go into an abuse situation in which it is not possible to prevent TR.

7.7 Impedance

A number of investigators have published articles demonstrating that the cell impedance may provide accurate information on its state. Based on this information, it is possible to estimate the SOC [71], the SOH [72] and even the internal temperature of the cell [73–79]. This makes it possible to improve the performance of the battery, extend its lifespan and guarantee its safety.

An electrochemical impedance spectroscopy is generally used to measure the cell impedance. This is a non-destructive technique that makes it possible to characterise the electrochemical behaviour of the batteries. To do so, the battery is subjected to sinusoidal current or voltage excitation, and the current or voltage response provided by each cell is measured. Based on the voltage and current ratio for each cell, it is possible to determine its complex impedance. Studies have demonstrated that the

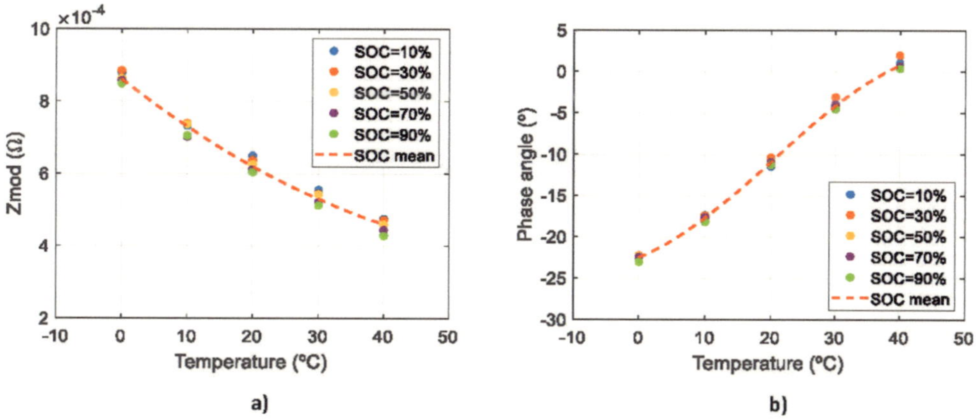

Figure 11.
Relationship between impedance and temperature. (a) Zmod vs. temperature; and (b) phase angle vs. temperature. Data obtained from [69].

impedance module [73], its real part [74], imaginary part [75], phase angle [76, 77] and intercept frequency [78, 79] can provide information on the internal state of the cells. **Figure 11** shows the results obtained from the EIS test data on 46 Ah pouch cells at different SOC and different temperatures [80]. On the one hand, **Figure 11a** shows the relationship between the cell impedance module (*Zmod*) and its temperature. By contrast, **Figure 11b** shows the relationship between the impedance phase and temperature. In both cases, the analysis was made on the impedance data for a frequency of 71.6 Hz, given that this is a frequency at which it has been demonstrated that there is less dependence on the cell SOC level. The study demonstrates that the impedance measurement can provide reliable information on the internal temperature of the cell.

The principal drawbacks of this detection method include the need for an impedance measurement system that does not interfere with battery performance. Moreover, as the excitation signal used for the batteries gives a low response, it is necessary to have extremely accurate sensors if the cell impedance is to be accurately estimated. In turn, given that the cell impedance varies in relation to the SOC and to the SOH, it would be necessary to perform a complete characterisation of the evolution of the impedance in the different scenarios to which LIBs may be exposed in their application.

7.8 Summary of the detection methods and research trends

Table 2 shows the main pros and cons of each detection variable. Despite the fact that no variable guarantees, in all cases, the detection of a TR phenomenon sufficiently in advance, the combination of the measurement of different variables would make the system safer and more reliable, significantly reducing the risk. However, due to the fact that the implementation of the measurement of new variables could increase the cost of the system, the pros and cons of each measurement must be considered in order to obtain a detection system that guarantees greater reliability with regard to safety and the detection of TR.

In this sense, in the roadmap of Battery 2030+ research initiative, R&D efforts are made towards the integration of smart functionalities in LIBs [81]. The integration of smart functionalities aims to enhance the lifetime and safety of batteries by injecting smart embedded sensing technologies and functionalities into the battery

Variable	Advantages	Disadvantages
Temperature	• Most indicative variable. • BMSs already measure it.	• It is not measured in all the cells. • Surface temperature is measured, not internal.
Voltage	• BMSs already measure it for all the cells. • Protects from overcharges and overdischarges.	• It does not anticipate TR detection in some abuses such us thermal until ISC occurs.
Current	• BMSs already measure it. • It can measure overcurrents and short circuits.	• In batteries with a large number of cells in series, a fault in one cell may not be detected.
Pressure	• Detects gas formation (internal measure). • Detects gas release (external measure)	• Internal measurement is complex. • Sealed module required.
Deformation	• Detects changes in volume due to gas formation.	• Hard to measure in cylindrical and prismatic. • Complex to integrate into large batteries.
Gas release	• Detects gas release (venting and TR). • Economical and feasible device	• Late detection if no venting is produced prior to TR, or if it is close to TR.
Impedance	• Early detection of TR. • Estimates cell internal temperature.	• New BMS functionality is required. • Complex measurement during operation.

Table 2.
Advantages and disadvantages of each detection variable.

cell, with the ultimate goal of integration into the BMS. SPARTACUS, SENSIBAT and INSTABAT are the three Research and Innovation actions granted under the H2020 topic 'Sensing functionalities for smart battery cell chemistries' with the aim of projects commissioned to develop affordable sensor solution to guarantee the safety of LIBs and to achieve an accurate monitoring of the state and the degradation phenomena. These projects propose alternative variables and measurement devices to ensure the safe operation of Li-ion batteries.

On the one hand, SPARTACUS aims to develop an affordable sensor solution based on acoustic sensors complemented by electrochemical impedance and temperature measurements. By holistic in-operando monitoring of these variables, SPARTACUS is developing a monitoring algorithm for the relevant battery states, with the state of safety (SoS) being one of them. By contrast, SENSIBAT chooses pressure, temperature and electrochemical impedance spectroscopy as indicative variables, developing the required sensing technology to be integrated in the battery cells. The goal of the project is to manufacture a module with the cells incorporating the designed sensors, thereby analysing the scaling-up of the technology, concluding the study with a cost-benefit analysis and a recycling study. Finally, INSTABAT relies on the concept 'Lab-on-a-cell', which is a multi-sensor platform that includes both physical sensors, such as optical fibre, a reference electrode or a photo-acousting sensor, and virtual sensors, with the thermal and electrochemical virtual sensors being the main proposal of the project. INSTABAT aims to enhance battery performance, reduce the required safety margins, improve the cell operation using data from the sensors and enable battery self-healing and second life usage.

The development of multi-variable and affordable sensing techniques able to combine the battery state estimation with the safety supervision is one of the most promising research trends related to the early detection of a TR.

8. Methods for the prevention and mitigation of TR

8.1 Cell safety

Protection at a cell level is primarily achieved through the use of electrical devices and subsystems that prevent excess temperature and pressure in the cell by opening the circuit, increasing the resistance or changing the chemical composition of the cell [41]. These devices include:

8.1.1 Thermal fuse

A thermal fuse is an alloy wire with a resistor and with thermal characteristics that allow it to melt when a given current flows or when exposed to excessive temperature. Once melted, it permanently disconnects the battery and needs to be replaced. Thermal fuses are used to protect against TR and their cut-off temperature is around 30–50°C above the maximum battery temperature in normal operation [21]. At high voltages, the disconnection of the fuse will not interrupt TR [82].

8.1.2 Positive temperature coefficient (PTC)

This is a resistor that increases its resistance as its temperature increases, so that it responds to a temperature increase by reducing the current and, therefore, the heat generated by the cell. At a normal temperature, its resistance is low. The principal purpose of PTC devices is to protect the batteries from external short circuits and also to provide protection in other conditions of electrical abuse [21].

8.1.3 Shutdown separator

The separator consists of a porous membrane intended for the purpose of allowing lithium ions to pass through it and electrically isolating the cell electrodes. As mentioned in Section 4, the most widely used materials for the separator are PE and PP. When used together, the protection achieved is known as a shutdown separator due to the different melting points of each element, 130°C for PE and 165°C for PP. When the PE melting point is reached and it melts, it fills the pores of the PP layer, preventing the lithium ions from flowing through and, therefore, shutting down the cell. This would make it possible to reduce the heat generated by the operation itself and to detect abnormal behaviour of the LIB. However, if the temperature continues to rise and reaches the melting point of the PP, then the separator loses its mechanical integrity and an ISC could occur. This protection must be designed to act at a temperature that is lower than the TR initiation temperature and with a shutdown temperature that is sufficiently below the temperature at which the separator loses its mechanical integrity [83].

8.1.4 Current interrupt device (CID)

Its main function is to protect the cell when the internal pressure increases above a maximum threshold. When the said pressure is reached in the cell interior, the electrical contact between the cathode and casing opens, interrupting the electric current and stopping the cell from operating. This is useful if the cause of the over-pressure is due to the current flowing through the cell. However, if this is not the cause, then it does not prevent the internal pressure from increasing, and it will be necessary for the safety vent to operate.

8.1.5 Safety vents

The cylindrical and prismatic cells generally incorporate this type of safety device, which opens when the pressure inside the cell reaches dangerous levels that could lead to an explosion or the rupture of the casing. By releasing the gases that were contributing to the internal pressure and temperature increase, cooling occurs due to the Joule-Thomson effect, and this could stop the temperature increase. However, it should be considered that, when the safety vent opens and releases the accumulated gases, atmospheric air (containing oxygen and moisture) also enters the cell and could react with lithium metal and electrolyte, causing ignition or an explosion [84].

8.2 Module safety

Module safety is achieved through the cell external management systems which control the operation of the LIB and monitor variables such as voltage, current and temperature in order to ensure that they remain within the limits of the SOA.

8.2.1 BMS

The BMS is the electronic system that controls the battery status and guarantees safe operation and optimal performance. For this purpose, it monitors variables such as the voltage, current and temperature of the cells and checks that these values stay within the safe operating ranges. It also performs a number of functions such as the balancing of the cell voltages, estimation of SOC and the control of the charge and discharge processes, among others.

Despite the fact that the BMS monitors the variables to ensure that they remain within the safe operating window, it is not able to prevent the onset of TR in all circumstances, in view of the accidents that continue to occur. However, the BMS is capable of detecting a TR and of directing the measures necessary for its mitigation and preventing its propagation. For this reason, there is a need to provide the BMS with new functions that make it possible to better estimate the battery status, as it is actually one of the aims of Battery 2030+ research initiative. Algorithms are currently being developed which, based on voltage, current and temperature, will provide information about the battery status, such as the internal temperature of the cell [85, 86], the triggering of an ISC [31, 87] and SOH [88]. Moreover, in combination with sensors that measure other variables, the detection of TR can be improved.

8.2.2 BTMS

The problems of excessive temperatures or local hot spots (affecting the uniformity of the battery) can be harmful to LIBs, affecting their performance, lifespan and safety. This problem is particularly exacerbated by the fact that, in the battery packs, the cells are placed very close together in order to take full advantage of the high energy and power density of the LIBs. For this reason, TR can propagate and cause the failure of the entire battery. It is therefore important to design a suitable thermal management system that makes it possible to maintain the temperature within its desired range and to guarantee the uniformity of the temperature in the cell. The BTMS is responsible for performing this function and for monitoring the temperature of the LIBs, ensuring that it remains within the SOA. The BTMS is not only a prevention measure, but can also mitigate the effects of a TR in the battery, by dissipating a large amount of the heat generated and obstructing the propagation of the phenomenon. The BTMS can be active or passive. On the one hand, an active system consumes extra energy to power the fans or pumps responsible for the circulation of a cooling fluid. Furthermore, the components of an active BTMS are integrated as a subsystem of the BMS which, in this case, is given the added function of regulating the cell heating or cooling. On the other hand, a passive system consists of surrounding the cells with layers of materials with the capacity to improve the dissipation of the heat generated, isolate one cell from another and even include certain properties such as flame retardancy, in order to prevent propagation and fire.

Despite the advantages that the BTMS offers, it does have certain disadvantages such as the higher cost and volume of the ESS, reducing its energy density. Moreover, it does not prevent the root causes of TR, given that it is only responsible for providing temperature control, which can get out of control if the generation of heat exceeds the BTMS dissipation capacity. For this reason, it must be accompanied by other devices such as the BMS, which monitors the state of the battery and measures other variables that permit the early detection of TR.

8.2.3 Mitigation and extinguishing systems

Once TR has been initiated in a cell, the mitigation strategy involves taking the measures that make it possible to reduce the dangers of the phenomenon and prevent its propagation. If the battery ignites as a result of the combustion of the flammable gases in contact with the oxygen in the air, the risks increase, given that a greater amount of heat is released, favouring propagation. As detailed in Section 6, for combustion to occur, the elements of the fire triangle must all be present and combined in the right proportions. If one of the three conditions is not met, then combustion will not take place. For this reason, investigations are underway regarding the introduction of inert gas dilutions in order to reduce the concentration of oxygen, as a method to prevent or extinguish the fire, given that these dilutions can reduce the flame temperature and increase the flammability limit [89]. For conventional hydrogen fuels, the limiting oxygen concentration (LOC) of their flame is usually above 12%, so they cannot maintain a flame at a lower oxygen concentration [90]. Nitrogen (N_2) and argon (Ar) are the gases that are most widely used to fight conventional fires. The advantage of inert gases is that they do not react with other gases that may be present, and they have proven to be effective in reducing the concentration of toxic gas emissions. Furthermore, the use of gaseous extinguishing agents makes it possible to occupy all possible spaces without affecting the functioning of the rest of the battery components. However, reducing the oxygen concentration to under 12% is not sufficient to eliminate the battery flame [90]. This may be due to the fact that, during TR, the internal temperature of the battery and the flammable gases released can exceed 800°C, so that these preheated fuel gases can extend the LOC and fuel flammability [89, 90].

Figure 12 collects a scheme of the safety strategies of detection, prevention and mitigation against TR commented in this chapter. The scheme allows to observe the level of implementation of any of the strategies and the data and control communication flows.

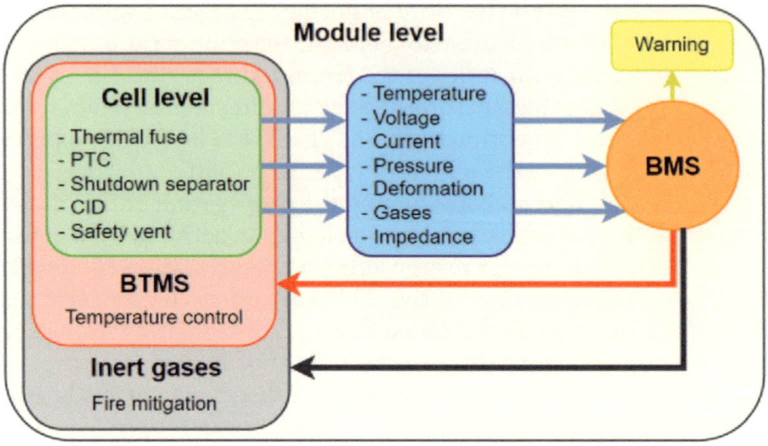

Figure 12.
LIBs safety scheme against TR.

9. Conclusions

The current scenario to combat climate change is providing a strong impetus to ESSs, directed at achieving greater integration of REs and the electrification of the automotive sector. LIBs are the most widely used technology in ESS, leading this transition towards the reduction of CO_2 emissions. While advances are continuing to be made with regard to the materials used in LIBs, in order to overcome the safety-related disadvantages and obtain higher energy densities, there is a need to gain in-depth knowledge of the behaviour of the cells outside their SOA and, in particular, all aspects surrounding the phenomenon of TR.

This chapter has addressed the problems of TR from a general point of view, analysing the main aspects that characterise it. The information provided makes it possible to gain a broad knowledge of this phenomenon and can prove useful as a guide and starting point for anyone interested in LIBs, in order to analyse aspects ranging from cell design and manufacture, battery assembly, right up to the end uses of these batteries.

This review of TR first looks at the causes that can trigger this phenomenon, it then goes on to analyse the phases occurring during the processes prior to the initiation of TR. By understanding the reactions and causes, the root of the problem can be tackled and the correct prevention measures can be taken. On the other hand, one of the particularities of thermal runaway is that it is influenced by a number of factors that may make prevention and detection more complicated. For this reason, a section has also been dedicated to understanding the factors affecting this phenomenon and the extent of the influence of each factor. Particular attention has been given to the variables that permit the early detection of TR and to mitigation strategies, given that, at present, TR in LIBs is a reality that we must learn to live with. Although there is no single variable that permits the early detection of TR, a combination of different variables guarantees greater accuracy in its detection. Once TR has been detected, the appropriate measures can be taken to prevent it from occurring or, should this prove impossible, to reduce its consequences. The BMS plays a crucial role in the safety of LIBs, as it is responsible for monitoring the main variables that determine cell behaviour and can even be adjusted to detect thermal runaway. Other safety measures such as BTMS for the improvement of the temperature control or inert gas dilutions to prevent battery fires are also helpful to fight TR.

Acknowledgements

This work is part of the projects PID2019-111262RB-I00, funded by MCIN/AEI/10.13039/501100011033/, STARDUST (774094), funded by European Union's Horizon 2020 research and innovation programme, and a FPI scholarship (PRE2020-095314), funded by Spanish Ministry of Science and Innovation. We would also like to acknowledge the support of Ingeteam R&D Europe.

Conflict of interest

The authors declare that there is no conflict of interest.

Author details

Iñaki Lalinde[1*], Alberto Berrueta[1], Juan José Valera[2], Joseba Arza[2], Pablo Sanchis[1] and Alfredo Ursúa[1]

1 Institute of Smart Cities, Department of Electrical, Electronic and Communications Engineering, Public University of Navarre (UPNA), Campus de Arrosadia, Pamplona, Spain

2 Ingeteam R&D Europe, Polígono Parque Tecnológico, Zamudio, Spain

*Address all correspondence to: inaki.lalinde@unavarra.es

References

[1] Millet L, Berrueta A, Bruch M, Reiners N, Vetter M. Extensive analysis of photovoltaic battery self-consumption: Evaluation through an innovative district case-study. Applied Physics Reviews. 2019;**6**(2). DOI: 10.1063/1.5049665

[2] CNESA. Energy Storage Industry White Paper 2021;**2021**(8610):2021

[3] Statista. Lithium-ion battery pack costs worldwide between 2011 and 2030. 2020. Available from: https://www.statista.com/statistics/883118/global-lithium-ion-battery-pack-costs/

[4] European Commission, Directorate-General for Internal Market, Industry, Entrepreneurship and SMEs, Blengini G, El Latunussa C, Eynard U, et al. Study on the EU's list of critical raw materials (2020): critical raw materials factsheets, Publications Office. 2020. DOI: 10.2873/92480. ISBN: 978-92-76-21053-5

[5] Feng X, Ouyang M, Liu X, Lu L, Xia Y, He X. Thermal runaway mechanism of lithium ion battery for electric vehicles: A review. Energy Storage Mater. 2018;**10**(May 2017):246-267. DOI: 10.1016/j.ensm.2017.05.013

[6] Feng X, Ren D, He X, Ouyang M. Mitigating thermal runaway of lithium-ion batteries. Joule. 2020;**4**(4):743-770. DOI: 10.1016/j.joule.2020.02.010

[7] Wang Q, Mao B, Stoliarov SI, Sun J. A review of lithium ion battery failure mechanisms and fire prevention strategies. Progress in Energy and Combustion Science. 2019;**73**:95-131. DOI: 10.1016/j.pecs.2019.03.002

[8] Miao Y, Liu L, Zhang Y, Tan Q, Li J. An overview of global power lithium-ion batteries and associated critical metal recycling. Journal of Hazardous Materials. 2022;**425**(November 2021):127900. DOI: 10.1016/j.jhazmat.2021.127900

[9] Qiu Y, Jiang F. A review on passive and active strategies of enhancing the safety of lithium-ion batteries. International Journal of Heat and Mass Transfer. 2022;**184**(xxxx):122288. DOI: 10.1016/j.ijheatmasstransfer.2021.122288

[10] Petzl M, Kasper M, Danzer MA. Lithium plating in a commercial lithium-ion battery—A low-temperature aging study. Journal of Power Sources. 2015;**275**:799-807. DOI: 10.1016/j.jpowsour.2014.11.065

[11] Vermeer W, Mouli GRC, Bauer P. A comprehensive review on the characteristics and modelling of Lithium-ion battery ageing. IEEE Transactions on Transportation Electrification. 2021;**8**(2):2205-2232. DOI: 10.1109/TTE.2021.3138357

[12] Mikolajczak C, White K, Kahn M, Long RT. Lithium-Ion Batteries Hazard and Use Assessment. SpringerBriefs in Fire. New York: Springer; 2011. XII, 115p. DOI: 10.1007/978-1-4614-3486-3. e-ISBN: 978-1-4614-3486-3

[13] Ruiz V, Pfrang A, Kriston A, Omar N, Van den Bossche P, Boon-Brett L. A review of international abuse testing standards and regulations for lithium ion batteries in electric and hybrid electric vehicles. Renewable and Sustainable Energy Reviews. 2018;**81**(May 2017):1427-1452. DOI: 10.1016/j.rser.2017.05.195

[14] Zhang X, Li Z, Luo L, Fan Y, Du Z. A review on thermal management of lithium-ion batteries for electric vehicles. Energy. 2022;**238**:121652. DOI: 10.1016/j.energy.2021.121652

[15] Berrueta A, Urtasun A, Ursúa A, Sanchis P. A comprehensive model for lithium-ion batteries: From the physical principles to an electrical model. Energy. 2018;**144**:286-300. DOI: 10.1016/j.energy.2017.11.154

[16] Petzl M, Danzer MA. Nondestructive detection, characterization, and quantification of lithium plating in commercial lithium-ion batteries. Journal of Power Sources. 2014;**254**:80-87. DOI: 10.1016/j.jpowsour.2013.12.060

[17] Han-Mei M. Side Reactions in Lithium-Ion Batteries. Berkeley: University of California; 2012

[18] Zhu J, Sun Z, Wei X, Dai H, Gu W. Experimental investigations of an AC pulse heating method for vehicular high power lithium-ion batteries at subzero temperatures. Journal of Power Sources. 2017;**367**:145-157. DOI: 10.1016/j.jpowsour.2017.09.063

[19] Soto A, Berrueta A, Oficialdegui I, Sanchis P, Ursúa A. Noninvasive aging analysis of lithium-ion batteries in extreme cold temperatures. IEEE Transactions on Industry Applications; IEEE. 2021;**58**(2):2400-2410. DOI: 10.1109/TIA.2021.3136808

[20] Soto A, Berrueta A, Sanchis P and Ursúa A. Analysis of the main battery characterization techniques and experimental comparison of commercial 18650 Li-ion cells. In: 2019 IEEE International Conference on Environment and Electrical Engineering and 2019 IEEE Industrial and Commercial Power Systems Europe (EEEIC/I&CPS Europe); 11-14 June 2019: Genova. New York: IEEE. p. 1-6. DOI: 10.1109/EEEIC.2019.8783862

[21] Balakrishnan PG, Ramesh R, Prem Kumar T. Safety mechanisms in lithium-ion batteries. Journal of Power Sources. 2006;**155**(2):401-414. DOI: 10.1016/j.jpowsour.2005.12.002

[22] Aurbach D et al. Design of electrolyte solutions for Li and Li-ion batteries: A review. Electrochimica Acta. 2004;**50**(2-3 SPEC. ISS):247-254. DOI: 10.1016/j.electacta.2004.01.090

[23] Lai X, Zheng Y, Zhou L, Gao W. Electrical behavior of overdischarge-induced internal short circuit in lithium-ion cells. Electrochimica Acta. 2018;**278**:245-254. DOI: 10.1016/j.electacta.2018.05.048

[24] Li HF, Gao JK, Zhang SL. Effect of overdischarge on swelling and recharge performance of lithium ion cells. Chinese Journal of Chemistry. 2008;**26**(9):1585-1588. DOI: 10.1002/cjoc.200890286

[25] Zhang L et al. Capacity fading mechanism during long-term cycling of over-discharged LiCoO2/mesocarbon microbeads battery. Journal of Power Sources. 2015;**293**:1006-1015. DOI: 10.1016/j.jpowsour.2015.06.040

[26] Zhao M, Dewald HD, Lemke FR, Staniewicz RJ. Electrochemical stability of graphite-coated copper in lithium-ion battery electrolytes. Journal of the Electrochemical Society. 2000;**147**(11):3983. DOI: 10.1149/1.1394007

[27] Maleki H, Howard JN. Effects of overdischarge on performance and thermal stability of a Li-ion cell. Journal of Power Sources. 2006;**160**(2 SPEC. ISS):1395-1402. DOI: 10.1016/j.jpowsour.2006.03.043

[28] Arora A, Medora NK, Livernois T, Swart J. Safety of Lithium-Ion Batteries for Hybrid Electric Vehicles. In: Electric and Hybrid Vehicles. Amsterdam: Elsevier B.V; 2010. p. 463-491. DOI: 10.1016/B978-0-444-53565-8.00018-X

[29] Kim KM, Jeong YS, Bang IC. Thermal analysis of lithium ion battery-equipped smartphone explosions. Engineering Science and Technology, an

International Journal. 2019;**22**(2):610-617. DOI: 10.1016/j.jestch.2018.12.008

[30] Berrueta A, Irigaray V, Sanchis P, Ursúa A. Lithium-ion battery model and experimental validation. In: 2015 17th European Conference on Power Electronics and Applications (EPE'15 ECCE-Europe); 08-10 September 2015: Geneva. New York: IEEE. p. 1-8. DOI: 10.1109/EPE.2015.7309337

[31] Lai X et al. Mechanism, modeling, detection, and prevention of the internal short circuit in lithium-ion batteries: Recent advances and perspectives. Energy Storage Materials. 2021;**35**(October 2020):470-499. DOI: 10.1016/j.ensm.2020.11.026

[32] Liu L et al. Internal short circuit evaluation and corresponding failure mode analysis for lithium-ion batteries. Journal of Energy Chemistry. 2021;**61**:269-280. DOI: 10.1016/j.jechem.2021.03.025

[33] Ren D et al. Investigating the relationship between internal short circuit and thermal runaway of lithium-ion batteries under thermal abuse condition. Energy Storage Materials. 2021;**34**(October 2020):563-573. DOI: 10.1016/j.ensm.2020.10.020

[34] Liu B et al. Safety issues and mechanisms of lithium-ion battery cell upon mechanical abusive loading: A review. Energy Storage Materials. 2020;**24**(June 2019):85-112. DOI: 10.1016/j.ensm.2019.06.036

[35] Chen Y et al. A review of lithium-ion battery safety concerns: The issues, strategies, and testing standards. Journal of Energy Chemistry. 2021;**59**:83-99. DOI: 10.1016/j.jechem.2020.10.017

[36] Lalinde I, Berrueta A, Sanchis P, Ursúa A. Applied method to model the thermal runaway of lithium-ion batteries. In: 2021 IEEE International Conference on Environment and Electrical Engineering and 2021 IEEE Industrial and Commercial Power Systems Europe (EEEIC/I&CPS Europe); 07-10 September 2021: Bari. New York: IEEE. p. 1-6. DOI: 10.1109/EEEIC/ICPSEUROPE51590.2021.9584560

[37] Ren D et al. A comparative investigation of aging effects on thermal runaway behavior of lithium-ion batteries. eTransportation. 2019;**2**. Article 100034, p. 1-13. DOI: 10.1016/j.etran.2019.100034

[38] Spotnitz R, Franklin J. Abuse behavior of high-power, lithium-ion cells. Journal of Power Sources. 2003;**113**(1):81-100. DOI: 10.1016/S0378-7753(02)00488-3

[39] Wang Q, Sun J, Yao X, Chen C. Thermal stability of LiPF6/EC + DEC electrolyte with charged electrodes for lithium ion batteries. Thermochimica Acta. 2005;**437**(1-2):12-16. DOI: 10.1016/j.tca.2005.06.010

[40] Richard MN, Dahn JR. Accelerating rate calorimetry studies of the effect of binder type on the thermal stability of a lithiated mesocarbon microbead material in electrolyte. Journal of Power Sources. 1999;**83**(1-2):71-74. DOI: 10.1016/S0378-7753(99)00260-8

[41] Wang Q, Ping P, Zhao X, Chu G, Sun J, Chen C. Thermal runaway caused fire and explosion of lithium ion battery. Journal of Power Sources. 2012;**208**:210-224. DOI: 10.1016/j.jpowsour.2012.02.038

[42] Kalhoff J, Eshetu GG, Bresser D, Passerini S. Safer electrolytes for lithium-ion batteries: State of the art and perspectives. ChemSusChem. 2015;**8**(13):2154-2175. DOI: 10.1002/cssc.201500284

[43] National Highway Traffic Safety Administration (NHTSA). Lithium-ion Battery Safety Issues for Electric and Plug-in Hybrid Vehicles (Report No. DOT HS 812 418). October, 2017

[44] Kawamura T, Kimura A, Egashira M, Okada S, Yamaki JI. Thermal stability of alkyl carbonate mixed-solvent electrolytes for lithium ion cells. Journal of Power Sources. 2002;**104**(2):260-264. DOI: 10.1016/S0378-7753(01)00960-0

[45] Campion CL, Li W, Lucht BL. Thermal decomposition of LiPF[sub 6]-based electrolytes for lithium-ion batteries. Journal of the Electrochemical Society. 2005;**152**(12):A2327. DOI: 10.1149/1.2083267

[46] Duan J et al. Building safe lithium-ion batteries for electric vehicles: A review. Electrochemical Energy Reviews. Springer Science and Business Media B.V. 2020;**3**(1):1-42. DOI: 10.1007/s41918-019-00060-4

[47] Pfrang A, Kriston A, Ruiz V, Lebedeva N, di Persio F. Safety of rechargeable energy storage systems with a focus on Li-ion technology. In: Emerging Nanotechnologies in Rechargeable Energy Storage Systems. Amsterdam: Elsevier Inc.; 2017. p. 253-290

[48] Son K et al. Comparative study of thermal runaway and cell failure of lab-scale Li-ion batteries using accelerating rate calorimetry. Journal of Industrial and Engineering Chemistry. 2020;**83**:247-251. DOI: 10.1016/j.jiec.2019.11.034

[49] Essl C, Golubkov AW, Fuchs A. Comparing different thermal runaway triggers for two automotive lithium-ion battery cell types. Journal of the Electrochemical Society. 2020;**167**(13):130542. DOI: 10.1149/1945-7111/abbe5a

[50] Börner M et al. Correlation of aging and thermal stability of commercial 18650-type lithium ion batteries. Journal of Power Sources. 2017;**342**:382-392. DOI: 10.1016/j.jpowsour.2016.12.041

[51] Golubkov AW et al. Thermal runaway of commercial 18650 Li-ion batteries with LFP and NCA cathodes—Impact of state of charge and overcharge. RSC Advances. 2015;**5**(70):57171-57186. DOI: 10.1039/c5ra05897j

[52] Zhao C, Sun J, Wang Q. Thermal runaway hazards investigation on 18650 lithium-ion battery using extended volume accelerating rate calorimeter. Journal of Energy Storage. 2020;**28**(January):101232. DOI: 10.1016/j.est.2020.101232

[53] Zhang Q, Liu T, Wang Q. Experimental study on the influence of different heating methods on thermal runaway of lithium-ion battery. Journal of Energy Storage. 2021;**42**(August):103063. DOI: 10.1016/j.est.2021.103063

[54] Fang J, Cai J, He X. Experimental study on the vertical thermal runaway propagation in cylindrical Lithium-ion batteries: Effects of spacing and state of charge. Applied Thermal Engineering. 2021;**197**(May):117399. DOI: 10.1016/j.applthermaleng.2021.117399

[55] Soto A et al. Integration of second-life battery packs for self-consumption applications: Analysis of a real experience. In: IEEE International Conference on Environment and Electrical Engineering and 2021 IEEE Industrial and Commercial Power Systems Europe (EEEIC/I&CPS Europe); 07-10 September 2021: Bari. New York: IEEE. 2021 p. 1-6. DOI: 10.1109/EEEIC/ICPSEUROPE51590.2021.9584809

[56] Braco E, San Martín I, Berrueta A, Sanchis P, Ursúa A. Experimental assessment of cycling ageing of lithium-ion second-life batteries from electric vehicles. Journal of Energy Storage.

2020;**32**(May):101695. DOI: 10.1016/j.est.2020.101695

[57] Braco E, San Martin I, Berrueta A, Sanchis P, Ursua A. Experimental assessment of first- and second-life electric vehicle batteries: Performance, capacity dispersion, and aging. IEEE Transactions on Industry Applications. Jul. 2021;**57**(4):4107-4117. DOI: 10.1109/TIA.2021.3075180

[58] Wu P et al. A study on thermal runaway of commercial lithium-ion cells: Influence of soc, cell chemistry and ageing status on safety performance. Lecture Notes in Electrical Engineering. 2019;**486**:619-627. DOI: 10.1007/978-981-10-8506-2_42

[59] Larsson F, Andersson P, Blomqvist P, Mellander BE. Toxic fluoride gas emissions from lithium-ion battery fires. Scientific Reports. 2017;**7**(1):1-13. DOI: 10.1038/s41598-017-09784-z

[60] Li W et al. Fire boundaries of lithium-ion cell eruption gases caused by thermal runaway. iScience. 2021;**24**(5):102401. DOI: 10.1016/j.isci.2021.102401

[61] Grandjean T, Barai A, Hosseinzadeh E, Guo Y, McGordon A, Marco J. Large format lithium ion pouch cell full thermal characterisation for improved electric vehicle thermal management. Journal of Power Sources. 2017;**359**:215-225. DOI: 10.1016/j.jpowsour.2017.05.016

[62] Carkhuff BG, Demirev PA, Srinivasan R. Impedance-based battery management system for safety monitoring of lithium-ion batteries. IEEE Transactions on Industrial Electronics. 2018;**65**(8):6497-6504. DOI: 10.1109/TIE.2017.2786199

[63] Zhu S et al. A novel embedded method for in-situ measuring internal multi-point temperatures of lithium ion batteries. Journal of Power Sources. 2020;**456**(July 2019):227981. DOI: 10.1016/j.jpowsour.2020.227981

[64] Jinasena A et al. Online internal temperature sensors in lithium-ion batteries: State-of-the-art and future trends. Frontiers in Chemical Engineering. 2022;**4**(February):1-22. DOI: 10.3389/fceng.2022.804704

[65] Pan Y, Feng X, Zhang M, Han X, Lu L, Ouyang M. Internal short circuit detection for lithium-ion battery pack with parallel-series hybrid connections. Journal of Cleaner Production. 2020;**255**:120277. DOI: 10.1016/j.jclepro.2020.120277

[66] Liao Z, Zhang S, Li K, Zhang G, Habetler TG. A survey of methods for monitoring and detecting thermal runaway of lithium-ion batteries. Journal of Power Sources. 2019;**436**(July 2019):226879. DOI: 10.1016/j.jpowsour.2019.226879

[67] Golubkov AW et al. Thermal runaway of large automotive Li-ion batteries. RSC Advances. 2018;**8**(70):40172-40186. DOI: 10.1039/C8RA06458J

[68] Koch S, Fill A, Birke KP. Comprehensive gas analysis on large scale automotive lithium-ion cells in thermal runaway. Journal of Power Sources. 2018;**398**(June):106-112. DOI: 10.1016/j.jpowsour.2018.07.051

[69] Cai T, Valecha P, Tran V, Engle B, Stefanopoulou A, Siegel J. Detection of Li-ion battery failure and venting with carbon dioxide sensors. eTransportation. 2021;**7**:100100. DOI: 10.1016/j.etran.2020.100100

[70] Essl C, Seifert L, Rabe M, Fuchs A. Early detection of failing automotive batteries using gas sensors. Batteries. 2021;**7**(2):1-26. DOI: 10.3390/batteries7020025

[71] Howey DA, Mitcheson PD, Member S, Yufit V, Offer GJ, Brandon NP. Motor controller excitation. IEEE Transactions on Vehicular Technology. 2014;**63**(6):2557-2566

[72] Kim J, Krüger L, Kowal J. On-line state-of-health estimation of Lithium-ion battery cells using frequency excitation. Journal of Energy Storage. 2020;**32**(September):101841. DOI: 10.1016/j.est.2020.101841

[73] Zhu JG, Sun ZC, Wei XZ, Dai HF. A new lithium-ion battery internal temperature on-line estimate method based on electrochemical impedance spectroscopy measurement. Journal of Power Sources. 2015;**274**:990-1004. DOI: 10.1016/j.jpowsour.2014.10.182

[74] Schmidt JP, Arnold S, Loges A, Werner D, Wetzel T, Ivers-Tiffée E. Measurement of the internal cell temperature via impedance: Evaluation and application of a new method. Journal of Power Sources. 2013;**243**:110-117. DOI: 10.1016/j.jpowsour.2013.06.013

[75] Spinner NS, Love CT, Rose-Pehrsson SL, Tuttle SG. Expanding the operational limits of the single-point impedance diagnostic for internal temperature monitoring of lithium-ion batteries. Electrochimica Acta. 2015;**174**:488-493. DOI: 10.1016/j.electacta.2015.06.003

[76] Srinivasan R, Carkhuff BG, Butler MH, Baisden AC. Instantaneous measurement of the internal temperature in lithium-ion rechargeable cells. Electrochimica Acta. 2011;**56**(17):6198-6204. DOI: 10.1016/j.electacta.2011.03.136

[77] Srinivasan R, Demirev PA, Carkhuff BG. Rapid monitoring of impedance phase shifts in lithium-ion batteries for hazard prevention. Journal of Power Sources. 2018;**405** (October):30-36. DOI: 10.1016/j.jpowsour.2018.10.014

[78] Raijmakers LHJ, Danilov DL, Van Lammeren JPM, Lammers MJG, Notten PHL. Sensorless battery temperature measurements based on electrochemical impedance spectroscopy. Journal of Power Sources. 2014;**247**:539-544. DOI: 10.1016/j.jpowsour.2013.09.005

[79] Raijmakers LHJ, Danilov DL, Van Lammeren JPM, Lammers TJG, Bergveld HJ, Notten PHL. Non-zero intercept frequency: An accurate method to determine the integral temperature of Li-ion batteries. IEEE Transactions on Industrial Electronics. 2016;**63**(5):3168-3178. DOI: 10.1109/TIE.2016.2516961

[80] Kilnk J, Orazov N, Grabow J, Benger R. Cell characteristics for 46Ah high power Kokam Nano pouch cell. Mendeley Data. 2020. DOI: 10.17632/g443f7cn7p.1

[81] Edström K, Dominko R, Fitchtner M, Otuszewski T, Perraud S, Punckt C, et al. BATTERY 2030+ Roadmap. 2020. Battery 2030+ [Internet]. Available from: https://battery2030.eu/wp-content/uploads/2022/07/BATTERY-2030-Roadmap_Revision_FINAL.pdf [Accessed: 2022-08-23]

[82] Chen YS, Hu CC, Li YY. The importance of heat evolution during the overcharge process and the protection mechanism of electrolyte additives for prismatic lithium ion batteries. Journal of Power Sources. 2008;**181**(1):69-73. DOI: 10.1016/j.jpowsour.2008.03.015

[83] Zhang SS. A review on the separators of liquid electrolyte Li-ion batteries. Journal of Power Sources. 2007;**164**(1):351-364. DOI: 10.1016/j.jpowsour.2006.10.065

[84] Belov D, Yang MH. Failure mechanism of Li-ion battery at

overcharge conditions. Journal of Solid State Electrochemistry. 2008;**12**(7-8):885-894. DOI: 10.1007/s10008-007-0449-3

[85] Anthony D, Wong D, Wetz D, Jain A. Non-invasive measurement of internal temperature of a cylindrical Li-ion cell during high-rate discharge. International Journal of Heat and Mass Transfer. 2017;**111**:223-231. DOI: 10.1016/j.ijheatmasstransfer.2017.03.095

[86] Sun L, Sun W, You F. Core temperature modelling and monitoring of lithium-ion battery in the presence of sensor bias. Applied Energy. 2020;**271**(March):115243. DOI: 10.1016/j.apenergy.2020.115243

[87] Zhang G, Wei X, Tang X, Zhu J, Chen S, Dai H. Internal short circuit mechanisms, experimental approaches and detection methods of lithium-ion batteries for electric vehicles: A review. Renewable and Sustainable Energy Reviews. 2021;**141**(January):110790. DOI: 10.1016/j.rser.2021.110790

[88] Lipu MSH et al. A review of state of health and remaining useful life estimation methods for lithium-ion battery in electric vehicles: Challenges and recommendations. Journal of Cleaner Production. 2018;**205**:115-133. DOI: 10.1016/j.jclepro.2018.09.065

[89] Li S, Zhang Y, Qiu X, Li B, Zhang H. Effects of inert dilution and preheating temperature on lean flammability limit of syngas. Energy and Fuels. 2014;**28**(5):3442-3452. DOI: 10.1021/ef500187s

[90] Weng J et al. Alleviation on battery thermal runaway propagation: Effects of oxygen level and dilution gas. Journal of Power Sources. 2021;**509**(June):230340. DOI: 10.1016/j.jpowsour.2021.230340

Chapter 4

Perspective Chapter: Design and Performance of Lithium-Ion Batteries for Achieving Electric Vehicle Takeoff, Flight, and Landing

Ruhul Amin, Nitin Muralidharan, Marm Dixit, Anand Parejiya, Rachid Essehli and Ilias Belharouak

Abstract

Today, the burgeoning drive towards global urbanization with over half the earth's population living in cities, has created major challenges with regards to intracity and intercity transit and mobility. This problem is compounded due to the fact that almost always urbanization and increase in standard of living drives individual automobile ownerships. Over 95% of automobiles are presently powered by some form of fossil fuel and as an unintended consequence, urban centers have also been centers for peak greenhouse gas emissions, a major contributor to global climate change. A revolutionary solution to this conundrum is flight capable electric automobiles or electric aerial vehicles that can tackle both urban mobility and climate change challenges. For such advanced electric platforms, energy storage and delivery component is the vital component towards achieving takeoff, flight, cruise, and landing. The requirements and duty cycle demands on the energy storage system is drastically different when compared to the performance metrics required for terrestrial electric vehicles. As the widely deployed lithium ion-based battery systems are often the primary go-to energy storage choice in electric vehicle related applications, it is imperative that performance metrics and specifications for such batteries towards areal electric vehicles need to be established. In this nascent field, there exists ample opportunities for battery material innovations, understanding degradation mechanism, battery design, development and deployment of battery control and management systems. Thus, this chapter comprehensively discusses battery requirements and identifies battery material chemistries suitable for handling aerial electric automobile duty cycles. The chapter also discusses the battery cell-level metrics pertaining to electrochemical, chemical, mechanical, and structural parameters. Furthermore, specific models for battery degradation, state of health (SOH), capacity and models for full cell performance and degradation are also discussed here. Finally, the chapter also discusses battery safety and future directions of batteries that would power these next generation urban electric aircrafts.

Keywords: EVTOL, battery, power density, energy density, payload

1. Introduction

By the year 2050, it is projected that ~70% of the global population will live in urban centers [1]. With the rapid rise in terrestrial automobiles (fossil fuel based) accompanying this urbanization drive, traffic congestions with high greenhouse gas emissions are unavoidable. Presently, on an average (according to a recent report based on 2019 statistics), American's waste ~99 h/year in traffic congestions [2], which is equivalent to a productivity loss worth ~88-billion USD. The revolutionary concept of urban air mobility (UAM) is increasingly being considered as a potential solution to mitigate this two-pronged conundrum. Recent reports suggest that even a small fraction of traffic diversion to UAM platforms can substantially reduce the vehicle fuel due to traffic congestions [3, 4]. With the recent successes and rapid advancements in electric vehicle technologies, the concept of electric Vehicle Take-off and Landing (eVTOL) has been garnering enormous research and development (R&D) consensus [5–7]. According to a recent Roland Berger report, there are over 95 reported eVTOL projects currently under various stages of development around the world and it even predicts that commercial passenger UAM routes are potentially possible by 2025 [8]. On the other hand, steady progress and advancements spanning several decades in battery technologies (specifically Lithium-ion batteries (LIBs)) has led to advanced microelectronics to power tools and grid load leveling to electric vehicles (EV). EVs powered by LIBs are already denting the carbon footprint of major cities worldwide. These LIB packs that power modern EVs are a product of continuous R&D efforts into the constituent material and design aspects driven by a positive feedback loop as a result of considerable data obtained on modern EVs. EVTOLs however, are a nascent technology and have unique operating profiles and duty cycles which places drastically different demands on LIB packs (with regards to calendar life, energy and power densities, safety, etc.,) that power them when compared to terrestrial EVs. Typical eVTOL batteries need to operate at both high C-rates (during take-off) and low C-rates during flight with intermittent high C-rate operation during cruise and areal maneuvering (see **Figure 1**), thus resulting in such batteries to have longer peak-power durations than EV batteries [9]. The operation of eVTOL batteries

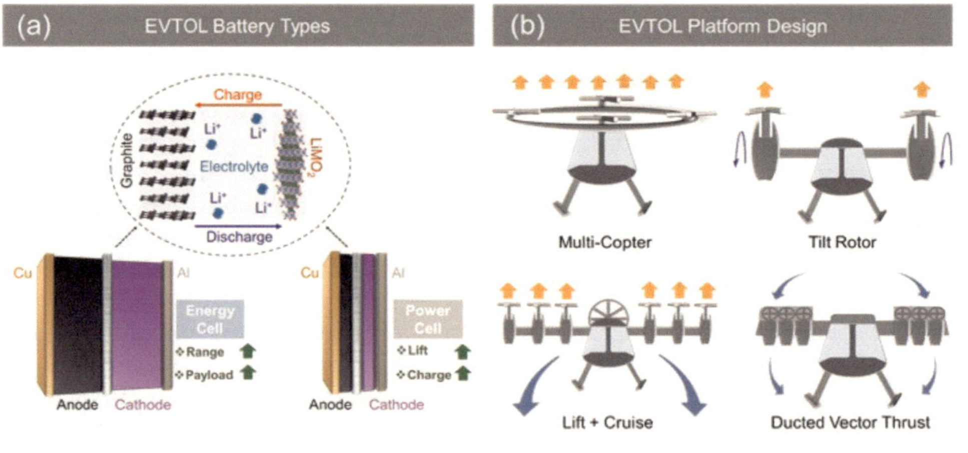

Figure 1.
(a) Designing of high-power density and high-energy density batteries for convenient eVTOL operation and (b) eVTOL design aspects for efficient energy utilization on boarding (multi-copter, tilt rotor, lift and cruise, and ducted vector thrust).

also vary depending on the distance traveled in a single flying time. It is also essential to fast charge to get sufficient energy in passenger-swapping gaps to ensure uninterrupted eVTOL operation in busy hours specifically in the case of public UAM platforms. In other times during lazy hours, the battery can be charged at lower C-rates or other charged batteries can be used to interchange with the exhausted one. The typical eVTOL design and its battery requirements metrics are displayed in **Figure 1** and detailed discussion is given later. It should be noted that such high utilization rates cause a critical challenge to battery cycle life as well as calendar life. Additionally, eVTOL batteries need to be designed in such a way that they should continue functioning and sustain the platform even after a safety incident occurs till a safe landing is achieved. In this chapter we highlight the materials selection for battery safety, fast charging capability, high -specific power, and -specific energy delivery for eVTOL applications. To meet the performance metrics for eVTOLs platforms, metrics such as energy density, power density, safety, cycle, and calendar life of battery systems need to be optimized and R & D focus should also be directed towards incorporating the energy dense and safe all solid-state batteries. Thus, this chapter highlights the opportunity for innovation in material developments, investigation and need to understand degradation mechanisms, development and deployment of battery control and management systems for eVTOL platforms. Specific objectives of this chapter include identification and down-selection strategies for battery chemistry selection for eVTOL applications, necessity, and strategies to implement predictive models for battery degradation and state-of-health (SOH), pathways to optimize advanced Li-ion materials and cell fabrication techniques for high energy and power density along with development of extensive electrochemical, chemical, mechanical, structural testing protocols towards eVTOL systems.

2. Battery metrics relevant for eVTOL applications

2.1 Specific energy/energy density

The energy density or specific energy of a lithium-ion battery depends on the capacity and operational cell voltage of the active materials, used inactive materials content and engineering design. The cell level energy density is defined by the (product of capacity and operational cell voltage)/total cell weight. Therefore, for batteries total system level energy density will be lower than the energy density at the cell level. The system level energy density can be enhanced if the use of inactive materials can be reduced. This can be achieved through the selection of appropriate materials combination. The energy landscape of various electrodes materials is depicted in **Figure 2a**. It can be inferred from the **Figure 2a** that selecting a high voltage cathode (positive electrode) and low voltage anode (negative electrode) with a high capacity will be the most favorable strategy to increase specific energy or energy density. Therefore, metallic lithium and silicon are excellent candidates to deploy as anodes for the next generation lithium-ion battery for eVTOL applications. The integration of high voltage cathode materials with the metallic lithium or silicon anode is not a straightforward process. The major barrier is to establish a suitable electrolyte which will be electrochemically stable at higher voltage and do not react with metallic lithium at lower voltage interval. The dendrite formation and shorting of the cell is the critical issue for lithium anode at high current rate either in liquid or solid electrolyte [10, 11]. It was anticipated that all solid-state battery (ASSB) can suppress the dendrite growth

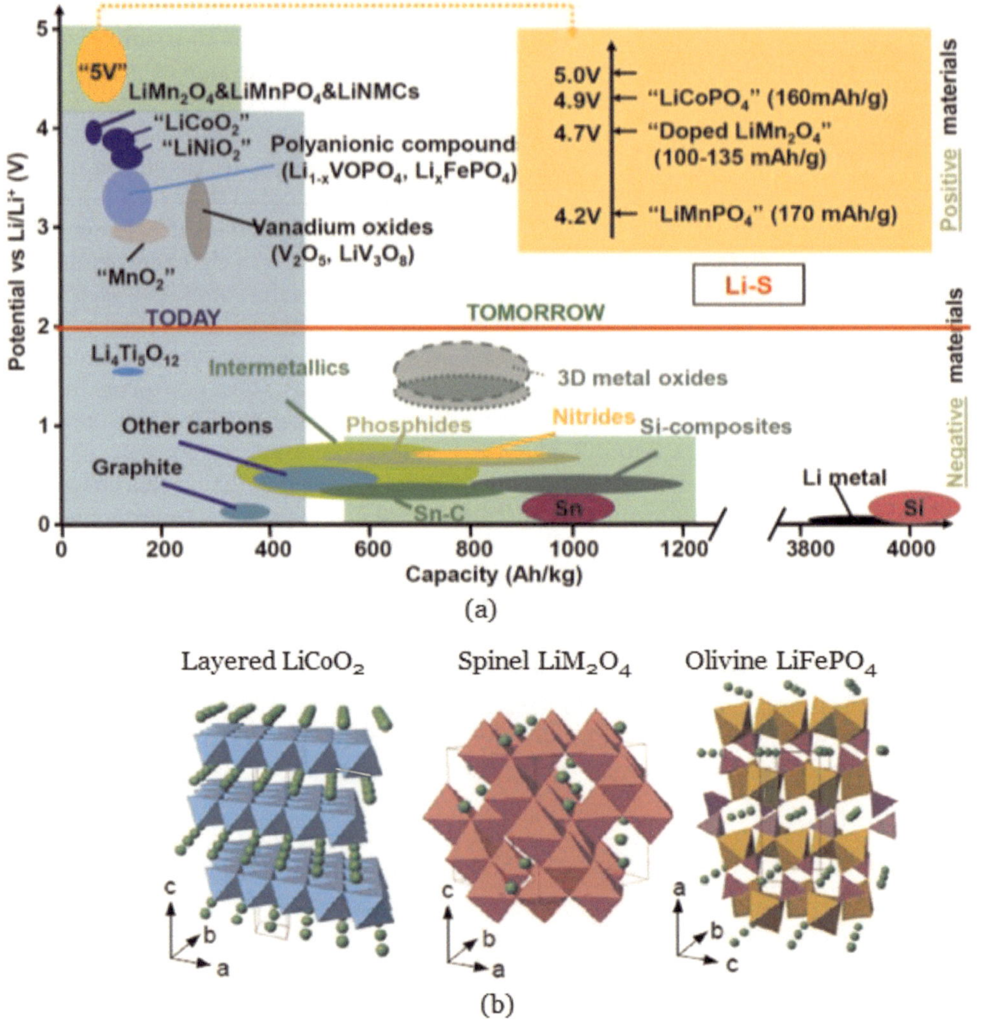

Figure 2.
(a) Energy landscape of various electrode materials: Capacity as a function of cell voltage against lithium electrode and (b) crystal structure of three main cathode materials, layered, spinel and olivine.

and protect the electrical cell shorting due to the high mechanical modulus solid electrolyte. However, the dendrite can grow through the solid electrode and can short circuit the battery cell. Because, ASSB has several interfaces such as anode | solid electrolyte (SE), cathode | SE, catholyte | active material, etc. These interface properties vary depending on the contact condition, energy states, type defects, and chemical/electrochemical stability. ASSB life and performance rely largely on these interfaces since dendrite formation, Li-depleted space-charge layer generation, and spatial variation in interfacial adhesion originate at the interfaces, which leads to battery failure. Stabilizing Li | SE interface is crucial for the development of high-energy density solid-state batteries. Current approaches in Li metal stabilization employ the energy and cost-intensive protocols that have a detrimental impact on the techno-economic feasibility of the ASSBs. Recently our group reported the facile, electrochemical protocol for improving the interfacial impedance and contact at the

Li | $Li_{6.25}Al_{0.25}La_3Zr_2O_{12}$ (LALZO) interface [10]. Noting that the implementation of a fraction of second short duration high voltage pulse to a poorly formed interface leads to a sustained improvement in contact impedance and lower overpotentials for electrodeposition and electro-dissolution. The good news is that our group invented the operando electrochemical dendrite healing process by using the current-voltage protocol to revive the battery cell from the failure state [11]. The problem of silicon is different than the metallic lithium. The volume expansion-contraction of lithiated and delithiated silicon is about 400% due to conversion/alloy formation reaction [12]. The silicon particle is disintegrated and electronically isolated during charge-discharge cycle and resulted sever capacity fade. However, the capacity fading issue is being mitigated partially by formation of silicon composite with carbon material. The various microstructural modification and coating design have been reported for silicon electrode which could not attract commercial viability due to difficulty of scalable processing [13].

Crystal structure, particle size and morphology, ionic and electrode conductivity of cathode materials play important role for the practical capacity and electrochemical performances of batteries at required current rate. The cathode materials can be classified into three main crystallographic structure (**Figure 2b**), (i) Layer structure (ii) spinel structure and (ii) olivine structure which are intercalation electrode and volume expansion-contraction is minimum during charge-discharge process. Each of the crystal structure has inherent advantages and disadvantages in terms of capacity, stability, cell voltage and electronic and ionic transport properties. Layer structure cathode exhibits higher practical capacity than spinel and olivine structure cathode. However, spinel and some olivine phases exhibit higher cell voltage than layer structure cathode. Details comparison is given in **Table 1**. Sulfur is only a high-capacity conversion type cathode material which has several short comes e. g. volume expansion-contraction, electronically insulator and poly sulfide dissolution in liquid electrolyte and shuttling impact [14]. These are the very critical issues which need to be addressed before commercial applications. The power density and energy density of a battery depend on the cell fabrication and battery design aspect as well as intrinsic properties of material. The energy cell or power cell can be fabricated deliberately by designing the coating thickness. The thicker electrode can provide higher energy density and lower power density for the same electrode material. On the other hand, thinner electrode design can produce higher power density but lower energy density. However, a specific electrode material can deliver a certain rang of practical power and energy density which depends on the ionic and electronic transport properties of the material and its inherent capacity and cell voltage. The elaborate discussion about power and energy density is given in session 2.2. The power density can also be tuned by playing around the particle size and morphology of the electrode materials for conventional lithium-ion battery. In all solid-state battery, electrolyte also play an important role for the power density. The smaller electrode particles are supposed to deliver higher power density compared to the bigger particles. Noting that the battery calendar and cycle life might be varied depending on the particle size. The smaller particle has high surface energy compared to the bigger particle which could originate parasitic reaction with other components of battery and induce shortening life of battery.

Specific energy is an important metric as the overall range or drive/flight distance of the eVTOL platform will depend on it. Higher the energy density of the constituent battery, longer the distance traveled. Moreover, high specific energies can also improve the payload of the eVTOL platform in addition to the range. The Breguet

		Voltage Window (V)	Nominal Capacity (mAh/g)	Current Cyclability	Worthiness for EVTOL Applications		Areas of Improvement
					Power Cell	Energy Cell	
Commercial Battery Chemistries	NMC Type ($LiNi_xMn_yCo_zO_2$, $x + y + z = 1$)	3–4.4	>200	Excellent (>1000 Cycles)	Compatible	Compatible	Electrode formulation and cell fabrication
	NCA Type ($LiNi_xCo_yAl_zO_2$, $x + y + z = 1$)	3–4.4	>200	Good (500–1000 Cycles)	Compatible	Compatible	Electrode formulation and cell fabrication
	Lithium Cobalt Oxide (LCO)	3–4.3	~180	Excellent (>1000 Cycles)	Compatible	Not Compatible	Structural stability and plugging it up into solid-state battery
	Lithium Iron Phosphate (LFP)	2.7–4.2	165	Excellent (>1000 Cycles)	Compatible	Not Compatible	Need to focus on higher voltage olivine phases
Battery Chemistries Close to Commercialization	Spinel Type (LNMO)	3–4.95	140	Excellent (>1000 Cycles)	Compatible	Not Compatible	Development of high voltage stable electrolyte
	Li-Silicon	0.05–1	3600	Average (100–500 Cycles)	Not Compatible	Compatible	Formation of good composite and minimization of first cycle capacity fade
	Nickel Rich & Cobalt Free	3–4.4	>200	Average (100–500 Cycles)	Compatible	Compatible	Cycling performance, particle size and morphology optimization and coating
Emerging Battery Chemistries	Li-Sulfur	1.8–2.8	~1600	Below Average (100–300 Cycles)	Not Compatible	Compatible	Translate to solid-state battery, formation of better composite electrode
	Lithium Rich Cathodes (LMR)	2–4.8	250	Below Average (100–300 Cycles)	Compatible	Compatible	Development of high voltage stable electrolyte and minimization of phase transition
	All-Solid-State Batteries	3–5	3862	Poor (100–300 Cycles)	Compatible	Compatible	Minimization of interface resistance and scalable fabrication

DOI: http://dx.doi.org/10.5772/intechopen.105477

		Voltage Window (V)	Nominal Capacity (mAh/g)	Current Cyclability	Worthiness for EVTOL Applications		Areas of Improvement
					Power Cell	Energy Cell	
Battery Chemistries in Lab scale R&D	Li Metal	3–4.4	3862	Average (100–500 Cycles)	Not Compatible	Compatible	Mitigation of dendrite effect
	Disordered rock salt cathodes	2–4	200–435	Average (100–500 Cycles)	Compatible	Compatible	Lower temperature synthesis and sintering.
	Li-Air	2–4	>2000	Very Poor (<100 Cycles)	Not Compatible	Compatible	Reduction of cell polarization and improving the cyclability
	Alloying anodes	0.05–1	>2000	Very Poor (<100 Cycles)	Not Compatible	Compatible	Minimization of impact of volume expansion-contraction
	$LiMPO_4$, M = Ni, V etc.	2.5–5	~200	Average (100–500 Cycles)	Compatible	Compatible	Tuning the transport properties and stabilization of nano particles

Table 1.
Comparison of electrochemical properties of various electrode materials and their compatibility with eVTOL applications and suggested the possible area of improvements.

range equation provides the relationship between the trip distance and the energy density requirement of the eVTOL battery [5, 15]:

$$R_{trip} = SE_{trip}\frac{L}{D}\frac{\eta_c}{g}\omega_{bat} \tag{1}$$

where R_{trip} denotes trip distance, SE_{trip} the specific energy consumed for the trip, and η_c the system efficiency, ω_{bat} is the battery weight fraction, and L/D is the lift-to drag ratio. It is evidenced from the above equation that the trip distance R_{trip} is directly proportional to SE_{trip} and ω_{bat} in the asymptotic limit of long range where energy consumption during takeoff and landing becomes negligible. The specific energy and specific power consumption of eVTOL batteries is very complex when compared to that of EV batteries. The traveling distance of an eVTOL platform also depends on the design and flying mechanism (flight apparatus configuration, for example: (a) multirotor, lift and cruise, tilt rotor and ducted thrust vectoring). Yang et al. identified three possible representative model designs of eVTOL aircraft configuration considering the parameter of energy consumption [9]. Contemporary designs have the lowest L/D ratio and represents the worst-case scenario for eVTOL configurations. The best design is expected to have a high L/D-ratio for efficient cruise as well as efficient hover which denotes the best-case scenario. The energy consumption should be considered sequentially for both safe landing and possible emergency diversions. Therefore, eVTOL batteries should not discharge to state of charge (SOC) below 10%, as voltage drops drastically in this region and leads to current spikes [16]. Additionally, eVTOL batteries should have reserve energy for balked landing or diversion to alternative locations in case of emergencies. As the battery ages (calendar/cycle life increase) the same distance cannot be travel with the same battery and the range would shrink further as the battery degrades. If a battery end of life is defined as one with 25% energy lost from its initial value, the nominal travel distance will drop accordingly. Potential eVTOL services can be intercity and intracity and thus aging batteries can be replaced to the intracity service from the intercity travel where short distances and reduced payload options can be considered. The maximum use of battery can be thus secured, and economic viability of eVTOL platforms can be achieved in such a way. EVTOL services are also envisioned as city taxi, airport shuttle, and intercity flights for passenger-UAM [8]. The first two scenarios are intracity hops with short trip distances and aging batteries with near end of life from the intercity vehicle can be installed for intra-city hops with short trip distances. Areal capacity of the battery cell can also be enhanced by fabricating thicker electrode architectures which could provide higher specific energy which can assist long distance travel, but the major challenge would be insufficient specific power for eVTOL operation as this imposes a limitation of fast charge/discharge rates.

2.2 Specific power/power density

The power density of a battery depends on two major factors. The intrinsic ionic and electronic transport properties of electrodes and electrolyte materials and design architecture of battery electrodes. The higher the transport properties of active materials, higher would be the power density. Some materials which have low conductivities may have other desirable properties in terms of eVTOL applications. The low conductivity materials need to be optimized by engineering specific to microstructures, particle morphologies and wiring with conductive additives. For instance, olivine structure ($LiFePO_4$) electrode materials exhibit lower ionic and electronic conductivity and one directional ionic diffusivity, however, boast stable cycling performances which would be vital for long calendar life. In such cases, nano particle with flake like morphologies or other desired architectures are highly recommended for high power battery development. While this is the case for the olivine structure, micron size particles with spherical morphologies are shown to deliver better performances in the case of layered electrode materials. Sulfur is the higher capacity insulator material which needs to be composited with a conductive carbon or polymer materials. The ionic conductivity and diffusivity of electrolyte materials are also very crucial for achieving higher power density in batteries. The ionic conductivity of various widely used electrolyte materials are shown in **Figure 3** as a function of operating temperature. This mode of comparison is valuable as operational temperature is also a critical parameter in battery performance for eVTOL platforms which would experience a wide range of thermal environments. The ionic conductivity of an electrolyte should lie within the red rectangle for convenient operation at a practical power density specific for eVTOL applications.

The relationship between power density and energy density of lithium-ion batteries is compared in **Figure 4**. It is evidenced from the **Figure 4** that power density is almost inversely proportional to the energy density i.e., higher the specific energy, lower is the specific power. Typically, eVTOL platforms cycle through different duty cycles with varying power demand and energy consumption while under operation. These phases typically include takeoff, hovering, climbing, cruising, descending, and landing. Yang et al., demonstrated a model of power and energy profile for a sample vehicle design for over an 80-km trip [9] which is redrawn in **Figure 5**. It is seen from **Figure 5** that takeoff and landing hovers have the highest power demand that

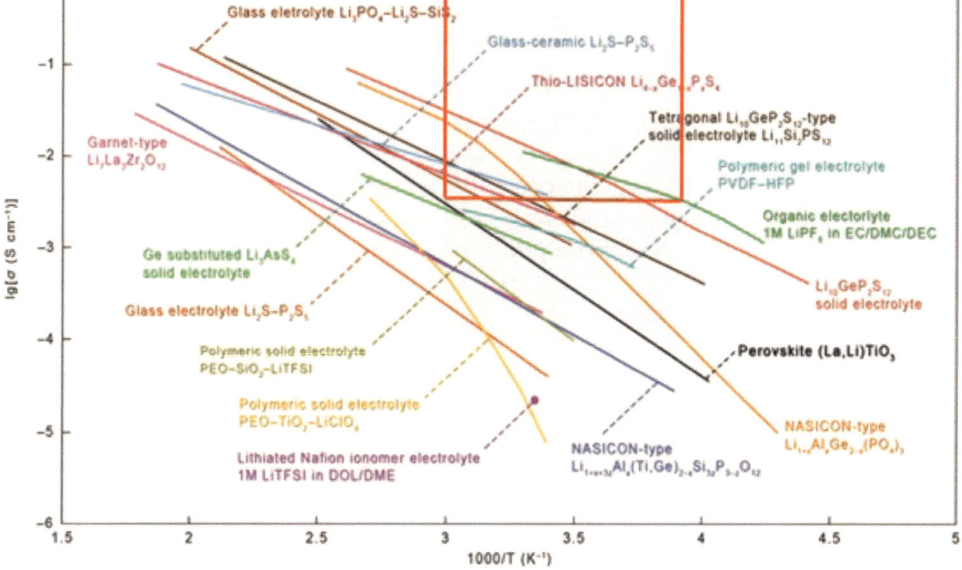

Figure 3.
Ionic conductivities of various electrolyte materials as function temperature.

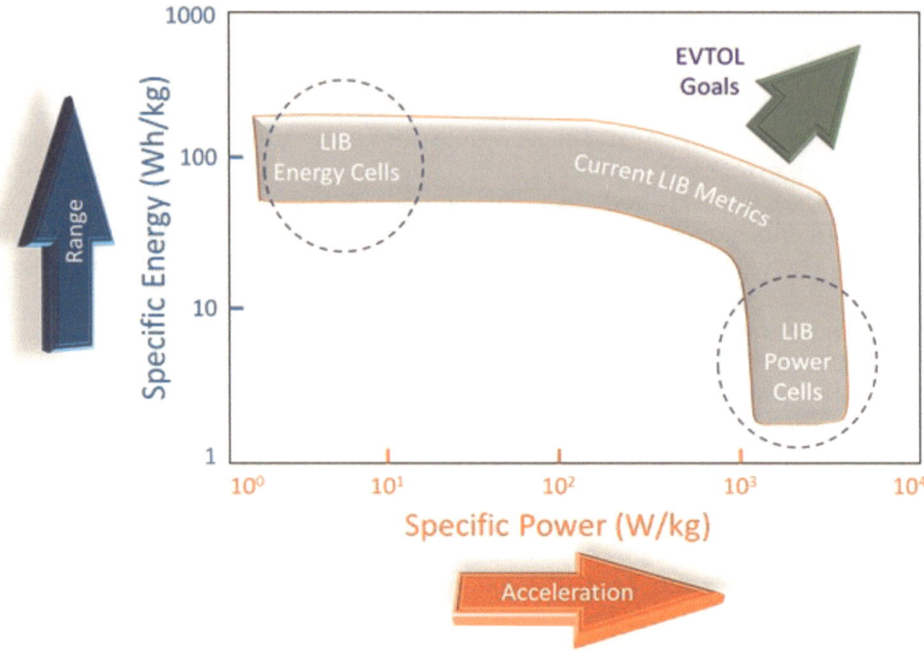

Figure 4.
Battery energy density as a function of power density consumption.

determines the battery's peak discharge rate, and the cruise power defines the battery's continuous discharge rate.

$$SP_{hover} = \frac{1}{\omega_{bat}} \frac{g}{\eta_h} \sqrt{\frac{\sigma}{2\rho_{air}}} \tag{2}$$

$$SP_{Cr} = \frac{1}{\omega_{bat}} \frac{g}{\eta_c} \frac{V_{Cr}}{L/D} \tag{3}$$

It is seen from the Eqs. (2) and (3) that, the specific power (SP) depends highly on battery weight fraction (ω_{bat}) and aircraft configuration—disk loading (σ) for hover-power and lift-to-drag (L/D) ratio for cruise power. It should be noted that aircraft design configuration determines the consumption of specific power while hovering and cruising [18]. For example, the wing-based eVTOL design demands less power efficiency during hovering and more power efficiency on cruising, opposite is applicable for multirotor based eVTOL [19, 20]. Thus, present designs demand a trade-off between hover and cruise efficiencies.

It has been previously discussed that eVTOL batteries operate at much higher C-rates than EV batteries [9]. The average C-rate of a typical EV battery is 0.3C in highway-driving and 0.1C in city driving, whereas a similar eVTOL battery averages around 1C over the 80 km trip and possibly even higher at hovering [9]. High specific power is required for both peak and continuous discharges which poses huge challenges to eVTOL batteries. As batteries face a power-energy trade-off an increased discharge power inevitably reduces the deliverable energy. Based on that the battery pack size should be optimized for a specific vehicle configuration to ensure sufficient energy output at the designed C-rates. On the other hand, both battery energy and power decrease substantially at freezing temperatures. Even for EVs, the cruise range

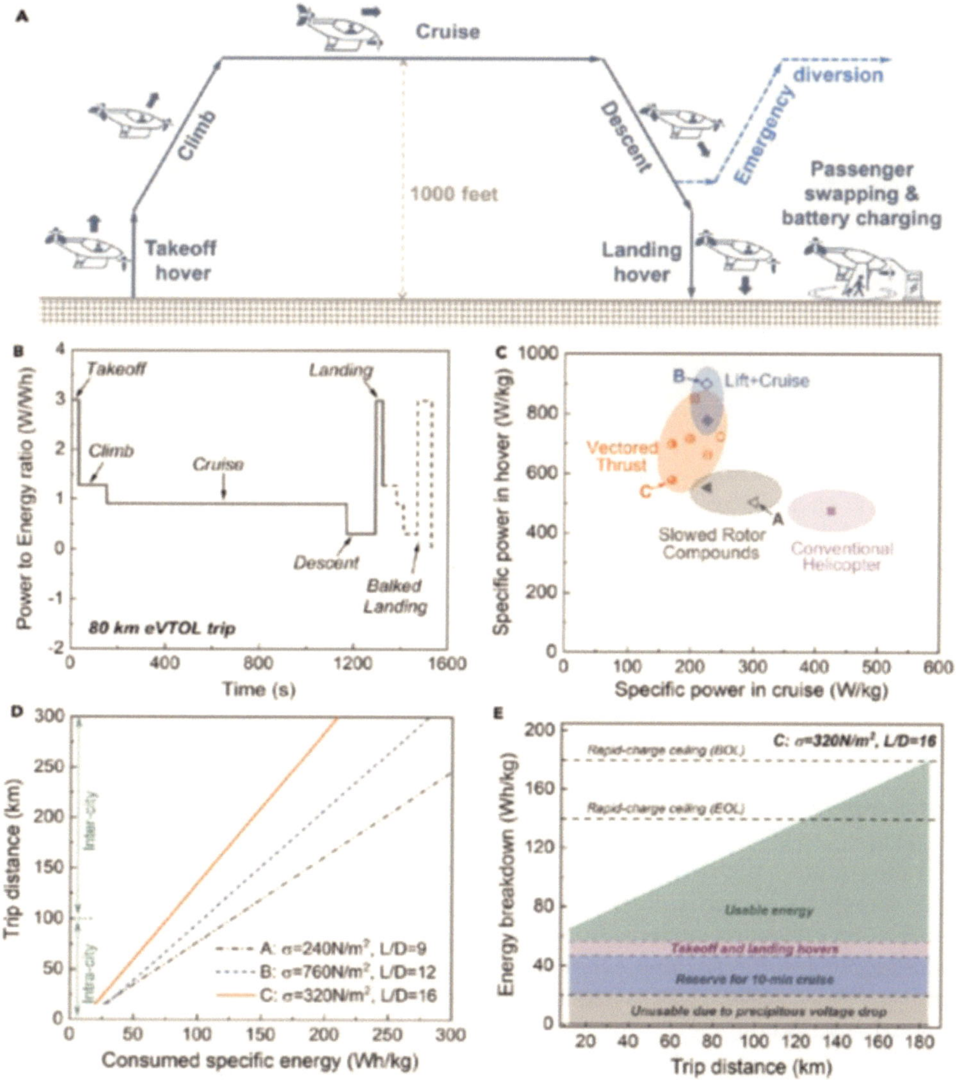

Figure 5.
The requirements of power and energy consumption in eVTOL battery for real world applications. Schematic illustration of a typical eVTOL trip (a) [17] representative battery power profile during an eVTOL trip and it depends on the vehicle design and operating pattern (B) [9] required battery specific power in hover versus in cruise for the aircraft configurations being pursued by the industry (C). Trip distance versus consumed specific energy for three representative aircraft configurations discussed in ref. [9] (D) and battery energy breakdown for eVTOL trips with specific design as mentioned in the ref. [9] (E).

could fall by <40% as the temperature drops from 24 to 7°C [21]. It is thus evident that the eVTOL batteries need to operate at higher C-rates and its operation would be more troublesome in cold weather when compared to conventional EVs.

2.3 eVTOL versus EV battery requirements

The degree of safety is an important platform specific parameter that varies based on the operational conditions of the battery. Battery for microelectronics has much

Operational conditions	EV battery	eVTOL battery	References
Driving/cruising C-rate	Need C/3 or C/10	Need 1C	[9, 12, 13]
Size of battery	Relatively small	Relatively big	
Power requirements	Relatively lower	Need higher	
Peak power duration	Typically, 10s	Typically lasting 30–120 s	
Battery lifetime	longer	shorter	
Charging frequency	less	more	

Table 2.
Comparison of EV and eVTOL batteries.

more different degree of safety than batteries for EVs. On the other hand, the eVTOL batteries have more rigorous safety requirements than EV batteries in all aspects. The high cruise power leads to a larger average discharge rate for eVTOL batteries. Thus, the specific energy of eVTOL batteries should be rated at a higher C-rate (e.g., 1C) than EV batteries (C/3, even C/10). The high fast-charging frequency and utilization rates make eVTOL batteries at present, more vulnerable than EV batteries which pose huge challenges to overall battery life. It should be noted that the duty cycle induced aging is more important than calendar aging for eVTOL batteries. The major differences between eVTOL and EV batteries are compared in the **Table 2**.

2.4 Need of fast charging

It is discussed in Eq. (1) that the eVTOL traveling range can be extended by elevating $\omega_{bat.}$ This is one way of increasing the travel distance. However, a higher ω_{bat} favorably reduces the required power density and fast charge rate. Lithium plating is big challenge for fast charging in energy-dense Li-ion cells due to the high cell polarization at an enhanced current. Energy dense cells needs thicker electrode in which fast charging and discharging event might be limited by ionic diffusion of electrolyte solution. Nonetheless, eVTOL needs larger batteries to ensure highly repeated operation during rush hours which have a small SOC consumption per trip. Therefore, the charge rate can be relaxed by recharging only partially in passenger-swapping gaps. However, larger the battery higher will be the costs and thus increasing payload demands. The aerospace batteries could cost 4.13 times more than automotive batteries on a US$/kWh basis as estimated by Uber [22].

Increasingω_{bat} is constrained by the payload, as:

$$m_{pay} = GTOM(1 - \omega_{bat} - \omega_{empty}) \tag{4}$$

where m_{pay} is the payload, GTOM the gross takeoff mass, and ω_{empty} the empty weight fraction. It should be noted that the ω_{empty} is an essential parameter of an aircraft. It is seen that a lower ω_{empty} is favorable but can only be achieved at a higher GTOM [21]. A constant ω_{empty} is typically presumed in eVTOL vehicle design and optimization for the sake of simplicities. It is obvious from the Eq. (4) that the raising ω_{bat} without sacrificing the payload requires increasing GTOM. It is shown in the above comparison in **Table 2** that the peak-power duration is typically 30–120 s for eVTOL takeoff or landing. It is estimated that annual effective flexible capacity (EFC) for eVTOL (~1600 EFCs) is much higher than EV battery (~45 EFCs) [9].

For fast-charging repetition frequency of eVTOL is much higher than EV battery [16] and within 5-min it should add energy from 50 to 62% SOC at the pack level, which is insufficient for eVTOL takeoff. The second fast-charging option can be adopted with a high-power electrodes design (thinner electrodes), that cannot offer sufficient specific energy for eVTOLs apparently. Furthermore, various steps have been taken to optimize charge algorithms which is simply able to manage the exclusion of Li plating without fundamentally relaxing the underlying electrochemical and transport limitations. The procedure can only incrementally enhance the fast-charging capability. A major technological challenge for eVTOL batteries is charging an adequate amount of energy in a very short time (very fast) without causing huge degradation to the batteries. In principle, Li plating can occur due to physicochemical processes at anode and electrolyte solution at higher C-rate which generate overpotential (cell polarizations). These are ion transport limitation in electrolyte solutions at high current, lithium intercalation reaction at interface between graphite-electrolyte and solid-state lithium-ion diffusion in graphite particles interior. It should be noted that ion transfer in the electrolyte is typically the limiting factor due to the thick and dense electrodes [22, 23] for high-energy LiBs. The possible solution would be enhancing electrolyte conductivity, diffusivity, and transference number [24–27] by tuning composition and proper formulation of the chemical components. However, it is not straight forward to improve one parameter without sacrificing other metrics due to the sheer complexity of these systems. Some solvents and salts can enhance ionic transport properties while electrolyte stability can be considerably compromised [28, 29]. As a result, battery cycle and calendar life are reduced compared to in normal operations. Another way would be aligning the electrode particles to reduce tortuosity [30–32]. It can be inferred from the above discussion that high energy density (thicker electrode) and big size batteries are not suitable for eVTOL operation on their own. The bigger batteries will cost on payload and high energy batteries limit fast charging. Therefore, energy density and size of battery needs to be critically assessed for eVTOL operation.

2.5 Charge pattern and cycle life

The cycle life depends on the depth of charge and discharge of battery during under operation and how fast it is charged and discharged. Lithium-ion batteries for eVTOL poses some unique challenges for fast energy storage and delivery on demand requirements. The electrical powertrains and batteries for eVTOL are needed to be coordinated for the energy and power delivery and consumption. The battery needs to supply huge energy instantly while the heavy gear of eVTOL needs to be hauled into the air depending on the payload of the eVTOL. There are some limitations to how much energy a battery can store and keep the weight light enough for the aircraft to achieve easy take off and flight. Developing an energy and power dense battery is one of the key challenges facing the electric aviation industry among other factors. In the case of public transit eVTOLs, it is worth noting that these need to run during rush hours more frequently (6–10 am and 4–8 pm per workday) in order to earn a profitable return on investment. Typically, we can envision a 5–10 min gap between two trips for passenger swapping in the peak hours. Thus, it is possible to charge the battery with enough energy needed for the next trip within this time. Based on the operational pattern it is assumed that eVTOL battery may need to charge several times in a day. For a short distance trip, depth of discharge might not cause a huge issue while for the relatively long-distance trip the depth of discharge expected to be huge to cover whole distance along with landing hovering. Thus, it can be concluded that fast charging is crucial for maximizing vehicle utilization rates in order to achieve eVTOL return on investment. Uber reported detailed analysis of earning revenues and predict eVTOL profit when compared with a typical car [7]. It should be noted that most charging events of eVTOL batteries are fast charging except in slow hour. The frequent charging and discharging will reduce the calendar life of battery and depth of charge and discharge will reduce both cycle and calendar life of battery. An alternative would be interchanging the discharged battery with charged battery in the station during passengers swapping for public eVTOL transit platforms. Furthermore, an interchangeable battery approach requires piles of reserve battery packs stored at each station which raises significant capital costs, safety, and space requirements. Making battery packs interchangeable would require standardization among battery and aircraft manufacturers, and market competition for consumers [33] and thus would be great challenge to overcome to establish a viable business.

2.6 eVTOL performance models and battery dataset generation

Operating conditions of eVTOL battery are different from other battery systems, and thus, performance models need to be developed for the whole calendar life of the battery system. Machine learning is a new modeling technique which provides more reliable performance prediction. However, the development of machine learning based models requires relevant battery performance datasets to train and test the models. There are only limited number of openly available datasets which poses a challenge to bring advanced machine-learning methods to battery performance prediction specific for eVTOL applications [34, 35]. In addition, there are no openly accessible datasets which follow the characteristic eVTOL duty cycle (Take-off, Cruise, Landing, Rest and Charging). Bills et al., [36] generated an experimental battery performance dataset specific to the power requirements of an eVTOL platform in order to fill this gap. Twenty-two cells were run in a variety of operating conditions that an eVTOL might experience during normal operation. The following charge-discharge cycling profiles were applied in the same generic format in all test cases:

a. Take-off: the cell was discharged at a high constant power for a period (around 120 s at least).

b. Cruise: the cell was discharged at a lower constant power for a longer duration,

c. Landing: the cell was discharged at high constant power (same rate as takeoff) for a slightly longer time,

d. Rest: the cell was allowed to rest until it cooled to a temperature of less than 27° C and again

e. Charging: the cell was charged using a constant current-constant voltage (CCCV) charging protocol,

f. Again Rest: The cell was allowed to rest until cell temperature reached 35°C, then allowed to rest 15 minutes further before beginning the next cycle.

After every 50 cycles, using the above eVTOL protocol, the capacity of each cell was tested and discharged at constant current until its voltage was 2.5 V. Also, the cells were cycled at temperatures exceeding 70°C throughout the followed charge-discharge protocols. Noting that during each test, temperature, voltage, current and discharge-charge capacity were recorded. The baseline profile was parameterized to test different conditions which could be encountered on an eVTOL operation mission. These test parameters, however, were not enough to implement good performance modeling. Furthermore, diffusion, transport and kinetics information of the cell components/interfaces were needed and were used in this model which cannot be directly obtained from these measurements. However, physically they correspond to processes which include electrochemical reactions, reaction kinetics, and transport. A subset of the parameters of the performance model was also chosen to capture the evolution of the performance model as the battery ages [15]. It should be mentioned that the parameters of battery models were very difficult to identify using local optimization methods [37]. Details about the model, method and discussion are reported in the Reference [36].

The reported aging model for discharge curves is shown in **Figure 6** which is important to model the life of the aircraft. Design decisions are thus made based on changes in the battery performance characteristics over time. It should be noted that the aging parameters must be allowed to evolve over the life of the battery in order to model the changes in performance characteristics over the battery's life.

Battery degradation model needs to be developed along with performance model. Bills et al., developed two degradation models [36]:

a. Mechanistic degradation model (MDM),

b. Universal battery degradation model (UBDM),

MDM can include contributions from solid electrolyte interface (SEI) growth, lithium plating, and active material loss [15, 25, 26]. These are three different

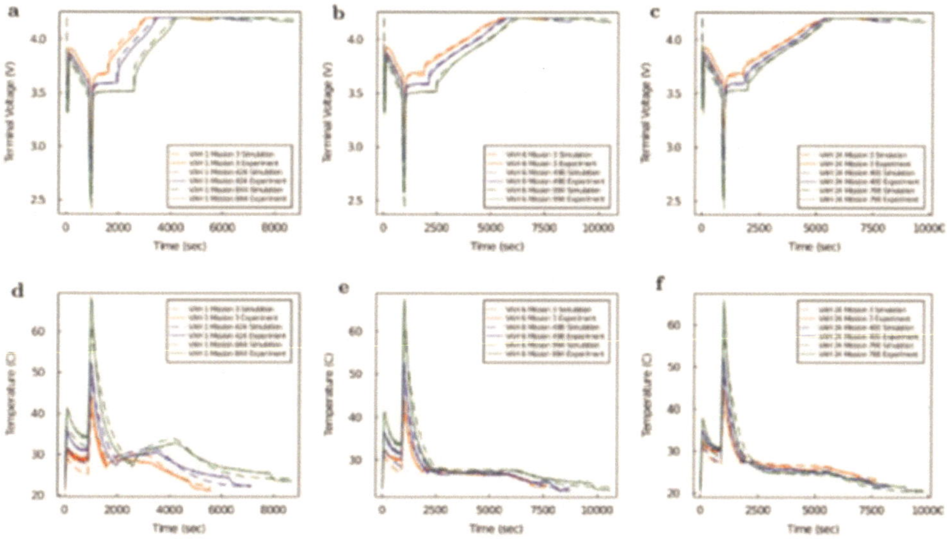

Figure 6.
Cellfit voltage and temperature predictions voltage (a-c) and temperature (d-f) pre-dictions along with experimental data generated at a sub-milliecond run time of Cellfit for 3 cells in the dataset [33].

degradation mechanisms and depend on the specific chemistry used in the battery. Some battery chemistries can exhibit severe impact, and some might be mild under operational conditions. Therefore, battery chemistry-based model should also be developed. UBDM approach can capture more complex effects of charge/discharge, active material loss as well as change in resistance [27, 30]. This was reported to be a differential education-based model, and thus any mechanistic model could be substituted for the currently used mechanistic model in the UBDM, to account for changes in chemistry, operating conditions, etc.

2.7 Safety of eVTOL battery

Safety is paramount for eVTOL batteries. Catastrophic failure of battery can occur due to various reasons. Fast charging is one of them which induces the plating of metallic lithium at the anode under extreme eVTOL duty cycle conditions. The plated Li metal is highly reactive and can trigger thermal runaway even without internal shorting [38, 39]. It should be noted that most eVTOL battery charging events are fast charging in nature and should be expected to be performed in all types of weather conditions. Thus plating-free, fast charging battery technologies are essential for secure eVTOL operation in extreme temperature conditions that an eVTOL platform experiences during its duty cycle. Additionally, high discharge rates of eVTOL batteries create severe challenges to battery thermal management systems (BTMS). Batteries have several resistive components which can cause heat on applying current. Typically, the battery heat generation rate is proportional to the square of the applied current. Therefore, heat generation of an eVTOL battery is higher than EV batteries as the discharge rate of eVTOL battery is generally considered to be higher (~1C) than a typical EV battery (C/3 for highway-driving and C/10 for city driving). Thus, a powerful BTMS is essential for the safe operation of eVTOL platforms. However, it is equally critical to limit BTMS mass to minimize the loss in pack-level-specific energy. The good news is that the rate of catastrophic failure for commercial LiBs has been reported to be only one in 40 million cells [40]. Statistics from Tesla has shown that for their electric vehicles, rate of fire incidents was only one in 205-million driving miles from 2012 to 2020 [41]. However, for eVTOL applications more rigorous battery design architectures need to be adopted to prevent all possible failures. Some novel features need be installed in the battery such as individual cell temperature monitoring abilities [42], self-healing/sealing electrodes and separators [43], etc. Furthermore, flammable liquid electrolyte in the conventional lithium-ion battery has been the critical component causing fire related hazards and thus alternatives such as all solid-state electrolytes can be considered when the technology matures. All solid-state-batteries (ASSB) are expected to provide more safety guarantees than conventional batteries. However, materials selection, development, and failure mode analysis are very critical for achieving sustainable operation of SSBs in eVTOL applications.

2.8 Confluence of EVTOL battery electrochemistry and mechanics

EVTOL platforms routinely experience a multitude of mechanical stimuli during take-off, cruise, and landing. These mechanical stimuli (vibrations) can be most often translated to the batteries that power these platforms. Such platforms can experience mechanical vibrations in the range of 10 Hz to 40 Hz [44]. Batteries that would power these EVTOL platforms need to be assessed for their airworthiness in terms of their mechanical robustness in addition to their electrochemical performance. Additionally, battery systems are inherently mechano-electrochemically coupled [45–47] and as a consequence, evaluation of battery material and component mechanical properties are essential for building EVTOL worthy battery packs. A multitude of mechanical properties pertaining to LIB systems for electric automobiles have been widely reported in literature. Commercial LIB cells used in electric vehicles which can be adapted for EVTOL platforms contain liquid electrolytes. These cells store a considerable amount of energy and unwarranted mechanical stimuli can lead to catastrophic short circuit events which can cause thermal runaways leading to structural compromise of the Battey pack which can cause damage to the EVTOL platform. Therefore, the individual structural components of the LIB cell have to be reassessed with regards to mechanical impulses experienced during EVTOL duty cycles. For example: the cell component that is expected to prevent short circuiting in LIB cells by preventing the direct contact between the two electrodes is the separator (usually made of a polymeric material). A good separator should be mechanically robust and should have sufficient puncture resistance without compromising electrochemical performance [48]. Apart from puncture resistance, the separator should also exhibit low thermal shrinkage. The electrodes should also be mechanically robust that there should be good adhesion between the electrodes and the active material coatings. High frequency mechanical vibrations experienced by the EVTOL platform during a duty cycle could compromise the adhesion between the active materials and the current collector. Specifically, during fully charged/discharged states as well as battery materials that have been subjected through hundreds of charge/discharge cycles could be adversely affected by high frequency mechanical vibrations. All solid-state batteries are also increasingly being considered for next generation EVTOL platforms that can operate at extreme temperature conditions. In such solid-state batteries, the inherent rigidity of the components when compared to their liquid-state counterparts, is expected to exacerbate the adverse effect of mechanical impulses experienced during EVTOL duty cycles. Moreover, current solid-state batteries require pressure application to achieve desired electrochemical performances. In such scenarios, a systematic and thorough R&D investigation of this phenomenon of mechano-electrochemical coupling is paramount for the development of mechanically robust batteries for EVTOL platforms.

3. All solid-state-battery (ASSB)

Based on the prior sections, we suggest that an ideal battery for eVTOL platform should be ultra-safe, delivering high energy density and preferably all solid-state (SSB) [49–56] that employs a high ionically conductive solid electrolyte (SE) which can operate in extreme temperatures (from −80–100°C) and at extreme environmental conditions like intense radiation and vibration (**Figure 7**). It is envisioned that such a battery can be deployed in applications pertaining to even outer space, upper altitudes, and harsh terrestrial environments. Different R&D efforts across the globe have been rigorously working towards achieving practical ASSBs. However, many challenges need to be overcome prior to successful commercialization. The hurdle is not only materials development but also strategizing scalable approaches.

Unlike conventional battery, high-capacity single crystalline cathode materials ($LiNi_xMn_yCo_zO_2$, NMC and xLi_2MnO_3, LMR) can be integrated with the best reported solid electrolyte (SE) combinations which demonstrate strong interfacial stability with both anode and cathode. Such cell assemblies when optimized, are expected to deliver

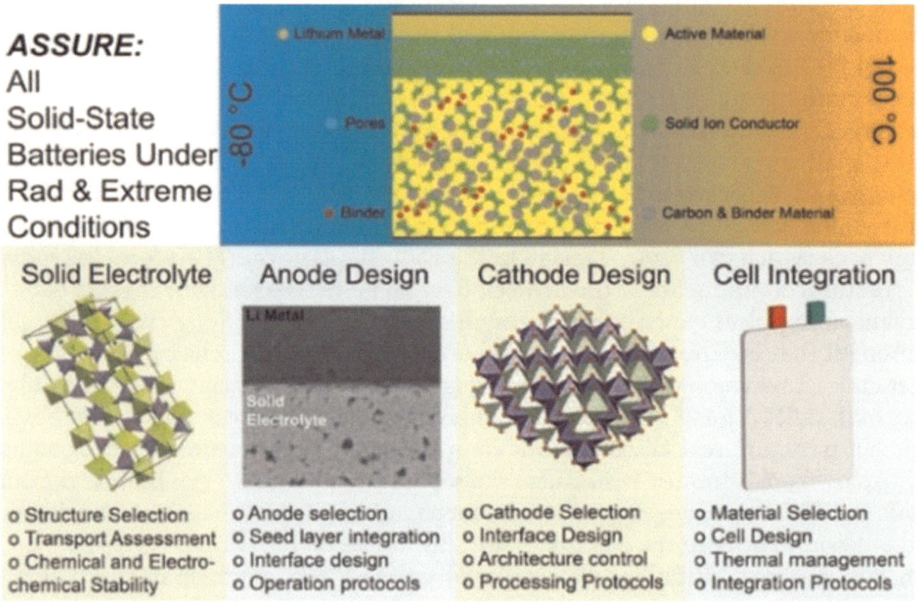

Figure 7.
Schematic diagram highlighting the design, focused area and application range, and aspects to be investigated for high energy density and high-power density all solid-state battery.

high energy and power densities. Ideal metrics include areal capacities >3 mAh cm^{-2}, demonstrating high Coulombic efficiencies over hundreds of cycles under extreme operating conditions. The ASSB has a transformational impact on secondary energy storage under harsh conditions as well as on specific battery deployments for conventional applications. At the cell level, inter and intra material resistances in solid-state batteries can significantly deteriorate the system performance in terms of capacity retention and coulombic efficiency. Identifying an upper bound of performance intrinsic to the material system can help in engineering the SSBs to mitigate the performance deficit in the meso-scale batteries. For this purpose, single crystal battery cathode is a promising approach and can be engineered using layered growth approaches (molecular beam epitaxy, atomic layer deposition, chemical vapor deposition, among others). Composite cathode designs can be tailored by utilizing material as well as engineering approaches. Material approaches can include high voltage cathode materials comprising of standard NMC622 material and Li- and Mn- rich cathode (LMR). LMR cathodes have high voltage and capacity but suffer from chemical degradation with conventional electrolytes which can be overcome with solid-state batteries. In order to mitigate intra-particle grain boundaries and transport limitations, single crystal/particle cathode materials can be a viable option for both NMC and LMR materials. In order to achieve high electrochemical performance, coating layers at the inter-particle interface between cathode active material and other components can be investigated that typically include LiA_xO_y (A = Nb, Ti, etc.), and LiX (X = halides). In addition to the identifying a stable cathode and solid electrolyte, R&D efforts should also focus on binder and electronic additives for battery stability in extreme conditions. Rigorous electrochemical testing (long term stability, capacity retention tests, GITT) should be performed on these batteries prior to deployment in eVTOL systems.

The economic viability is extremely important for the commercial deployment of EVTOL. It is a great challenge to transport people and cargo economically and efficiently in short and long-distance and is generated revenue by moving their luggage and freight. The cost and revenue are important but also very complex for EVTOL. The battery price, calendar and cycle life, and design cannot give a complete picture of cost and revenue. The payload, range and hours flying per day, and design of aircraft are key metrics. For instants, aircraft volume is a big concern because most flights are volume-limited rather than weight-limited. The batteries need to be carried within the wings as is fuel on current airliners, otherwise, the size of the aircraft must be increased to maintain payload volume. There are other direct costs including ownership, fuel, maintenance, and crew, airborne electrical equipment and gas turbines now cost about the same per kilowatt. All of these aspects must be considered when generating the cost rate or how much energy is consumed per passenger mile. The cost per kilowatt-hour is driven by the degradation of the cell which refers to the effect the charging itself does to degrade the use of the battery. Details cost analysis and economic comparison between electric aircraft and conventional airliners is reported in nature recently [57].

4. Future outlook

i. In ideal scenario, battery life is expected to last as long as vehicle life is sustained. Similar to terrestrial EVs, eVTOL platforms are also typically designed for achieving tens of years in service. However, design for available batteries would be based on the near term of 15 years at best. Additionally, the maximum range of an eVTOL vehicle is essentially proportional to the product of battery weight and specific energy as discussed in this chapter. Therefore, in future, R&D efforts have to be directed towards innovative battery designs and chemistry specific for eVTOL platforms. Specifically, enhancements to improve the specific energy of the battery would be ideal to achieve a larger payload while minimizing battery size for both short-range intracity commutes as well as for long-range intercity trips.

ii. With the conventional LIB chemistries, increase the areal loadings of active materials is the most practical approach to enhance specific energy. However, this approach creates a critical barrier for achieving fast charging capability. Yang et al., [9] previously reported that the battery cells (~215 Wh/kg) could be charged with 6C rates, enabling 5 min charging for 80 km trips. This implies that eVTOL batteries can be optimized based on their specific duty cycles. For example, a private eVTOL platform would require a specific type of battery optimized to serve a specific duty cycle while a public transit based eVTOL platform.

iii. Based on our discussions, it can be inferred that high areal capacities would reduce the charge rate further thus limiting eVTOL operation at rush hour which may help in achieving a better return on investment. Therefore, it is essential to develop advanced battery technologies and protocols that can safely achieve fast charging of ultrahigh-energy density lithium-ion batteries.

iv. The high utilization rates of eVTOL battery requires higher cycle life sustainability. Enhancing the specific energy and improvement of fast-charging capability could assist the cycle life longevity. The battery functionality (payload minimization) and performance are always a tradeoff. Presently, high voltage Ni-rich oxide cathodes with high-capacity silicon or Li-metal anodes are a prospective avenue to increase specific energy, however, such combinations when subjected to fast charging (potential eVTOL duty cycle requirements) can induce significant volume expansion of silicon anode, or formation and propagation of Li dendrites thereby resulting in either catastrophic failure or low cycle life expectancy. Nevertheless, significant breakthrough is needed in fast charging and cycle life advancements for new generation battery technologies to be viable for eVTOL applications.

v. Another way to increase the battery performance is to operate at higher temperature. Elevating the operational temperature can lower the cooling demand by an order of magnitude because of reduced heat generation rates and higher temperature differences between cells and ambient atmosphere [58, 59]. Elevated temperature operation can also improve battery power capability due to fast charging possibilities. Our opinion is that battery components with thermally modulated structures enables outstanding battery performance at low ambient temperatures, while the use of stable materials offers high safety and durability at high temperatures. This synergistic combination can achieve durable, safe, and high-performance, eVTOL operation even at different atmospheric conditions and thermal cycles.

Acknowledgements

This manuscript has been supported by Oak Ridge Nation Laboratory (ORNL) managed by UT-Battelle, LLC, under contract DE-AC05-00OR22725 with the US Department of Energy (DOE) and the Laboratory Directed Research and Development (LDRD) Program at Oak Ridge National Laboratory.

Notice

Author details

Ruhul Amin[1*], Nitin Muralidharan[1], Marm Dixit[1], Anand Parejiya[1,2], Rachid Essehli[1] and Ilias Belharouak[1]

1 Electrification and Energy Infrastructures Division, Oak Ridge National Laboratory, Oak Ridge, TN, USA

2 University of Tennessee, Bredesen Center for Interdisciplinary Research and Graduate Education, Knoxville, TN, USA

*Address all correspondence to: aminr@ornl.gov

References

[1] United Nations. World Urbanization Prospects, (2018 revision). 2018. https://population.un.org/wup/Publications/Files/WUP2018-Report.pdf

[2] INRIX. Global Traffic Scorecard. 2019. https://inrix.com/scorecard/

[3] Goyal R. Urban Air Mobility (UAM) Market Study (Technical Report HQ-E-DAA-TN65181, (NASA)). 2018. https://ntrs.nasa.gov/citations/20190001472

[4] Lin Z, Xie F, Ou S. Modeling the external effects of air taxis in reducing the energy consumption of road traffic. Transportation Research Record. 2020; **2674**:176-187

[5] Kasliwal A, Furbush NJ, Gawron JH, McBride JR, Wallington TJ, De Kleine RD, et al. Role of flying cars in sustainable mobility. Nature Communications. 2019;**10**:1555

[6] Sutherland BR. Flying cars for green transportation. Joule. 2019;**3**: 1187-1189

[7] Holden J, and Goel N. Fast-Forwarding to a Future of on-Demand Urban air Transportation ((Uber Elevate)). 2016. https://www.uber.com/elevate.pdf

[8] Harder M, Baur S, Kopera S, Schonberg T, Hasenberg JP. The High-Flying Industry: Urban Air Mobility Takes Off (Roland Berger). 2020. Available from: https://www.rolandberger.com/en/Insights/Publications/The-high-flying-industry-Urban-Air-Mobility-takes-off.html

[9] Yang X-G, Liu T, Ge S, Rountree E, Wang C-Y. Challenges and key requirements of batteries for electric vertical takeoff. Joule. 2021;**5**:1644-1659

[10] Parejiya A, Amin R, Essehli R, Wood DL, III, Belharouak I. Electrochemical healing of dendrites in garnet-based solid electrolytes. ACS Energy Letters. 2021;**5**: 3368-3373

[11] Parejiya A, Amin R, Dixit MB, Essehli R, Jafta C, Wood DL, et al. Improving contact impedance via electrochemical pulses applied to lithium–solid electrolyte interface in solid-state batteries. ACS Energy Letters. 2021;**6**:3669

[12] Szczech JR, Jin S. Nanostructured silicon for high capacity lithium battery anodes. Energy & Environmental Science. 2011;**4**(1):56-72

[13] Xu K, Liu X, Guan K, Yu Y, Lei W, Zhang S, et al. Research Progress on coating structure of silicon anode materials for Lithium-ion batteries. ChemSusChem. 2021;**14**(23):5135-5160

[14] Zhang SS. Liquid electrolyte lithium/sulfur battery: Fundamental chemistry, problems, and solutions. Journal of Power Sources. 2013;**231**: 153-162

[15] Patterson MD, German BJ, Moore MD. Performance analysis and design of on-demand electric aircraft concepts. In: 12th AIAA Aviation Technology, Integration, and Operations (ATIO) Conference and 14th AIAA/ISSMO Multidisciplinary Analysis and Optimization Conference. 2012. Available from: https://arc.aiaa.org/doi/10.2514/6.2012-5474

[16] McDonald R, German B. eVTOL Stored Energy Overview (Uber Elevate Summit 2017). 2017. Available from: https://www.uber.com/us/en/elevate/summit/2017/

[17] Airbus. Urban air Mobility Iconography. 2018. https://acubed.airbus.com/blog/airbus-utm/urban-air-mobility-iconography/

[18] Moore M. Making eVTOL Real. Uber Elevate Summit. 2019. Available from: https://evtol.news/news/uber-elevate-summit-2019-agenda

[19] Moore M. eVTOL cost drivers (Uber Elevate Summit 2019). 2019

[20] Raymer D. Aircraft Design: a Conceptual Approach (American Institute of Aeronautics and Astronautics). 2012

[21] Moloughney T. Tesla V3 supercharger test: we find out exactly how fast it really is. 2020. https://insideevs.com/news/427836/tesla-model-3-v3-supercharging-test/

[22] Ahmed S, Bloom I, Jansen AN, Tanim T, Dufek EJ, Pesaran A, et al. Enabling fast charging – A battery technology gap assessment. Journal of Power Sources. 2017;**367**:250-262

[23] Colclasure AM, Tanim TR, Jansen AN, Trask SE, Dunlop AR, Polzin BJ, et al. Electrode scale and electrolyte transport effects on extreme fast charging of lithium-ion cells. Electrochimica Acta. 2020;**337**:135854

[24] Zhang SS. Reformulation of electrolyte for enhanced fast-charge capability of Li-ion battery. Journal of the Electrochemical Society. 2020;**167**: 060527

[25] Logan ER, Dahn JR. Electrolyte design for fast-charging Li-ion batteries. Journal of Trends in Chemistry. 2020;**2**: 354-366

[26] Zhang X, Zou L, Xu Y, Cao X, Engelhard MH, Matthews BE, et al. Advanced electrolytes for fast-charging high-voltage lithium-ion batteries in wide-temperature range. Advanced Energy Materials. 2020;**10**:2000368

[27] Du Z, Wood DL, Belharouak I. Enabling fast charging of high energy density Li-ion cells with high lithium ion transport electrolytes. Electrochemistry Communications. 2019;**103**:109-113

[28] Wu X, Liu T, Bai Y, Feng X, Rahman MM, Sun C-J, et al. Effects of solvent formulations in electrolytes on fast charging of Li-ion cells. Electrochimica Acta. 2020;**353**:136453

[29] Logan ER, Hall DS, Cormier MME, Taskovic T, Bauer M, Hamam I, et al. Ester-based electrolytes for fast charging of energy dense lithium-ion batteries. Journal of Physical Chemistry C. 2020; **124**:12269-12280

[30] Shi B, Shang Y, Pei Y, Pei S, Wang L, Heider D, et al. Low tortuous, highly conductive, and high-areal-capacity battery electrodes enabled by through-thickness aligned carbon fiber framework. Nano Letters. 2020;**20**:5504-5512

[31] Chen KH, Namkoong MJ, Goel V, Yang C, Kazemiabnavi S, Mortuza SM, et al. Efficient fast-charging of lithium-ion batteries enabled by laser-patterned three-dimensional graphite anode architectures. Journal of Power Sources. 2020;**471**:228475

[32] Vishnugopi BS, Verma A, Mukherjee PP. Fast charging of lithium-ion batteries via electrode engineering. Journal of the Electrochemical Society. 2020;**167**: 090508

[33] Ahmad F, Saad Alam M, Saad Alsaidan I, Shariff SM. Battery swapping station for electric vehicles:

Opportunities and challenges. IET Smart Grid. 2020;**3**:280-286

[34] Severson KA et al. Data-driven prediction of battery cycle life before capacity degradation. Nature Energy. 2019;**4**:383-391. DOI: 10.1038/s41560-019-0356-8

[35] Howey DA. Tools for battery health diagnostics and prediction. Electrochemical Society Interface. 2019;**28**:55-56. DOI: 10.1149/2.f06191if

[36] Bills A, Sripad S, Fredericks WL, Guttenberg M, Charles D, Frank E, et al. Universal battery performance and degradation model for electric aircraft. rXiv:2008.01527v2 [physics.app-ph]. 16 Mar 2021:1-30. DOI: 10.26434/chemrxiv.12616169.v1

[37] Forman JC, Moura SJ, Stein JL, Fathy HK. Genetic identification and fisher identi-fiability analysis of the Doyle–fuller–Newman model from experimental cycling of a LiFePO4 cell. Journal of Power Sources. 2012;**210**:263-275. DOI: 10.1016/j.jpowsour.2012.03.009

[38] Waldmann T, Hogg B-I, Wohlfahrt-Mehrens M. Li plating as unwanted side reaction in commercial Li-ion cells – A review. Journal of Power Sources. 2018;**384**:107-124

[39] Li Y, Feng X, Ren D, Ouyang M, Lu L, Han X. Thermal runaway triggered by plated lithium on the anode after fast charging. ACS Applied Materials & Interfaces. 2019;**11**:46839-46850

[40] Doughty DH, Roth EP. A general discussion of Li ion battery safety. Electrochemical Society Interface. 2012;**21**:37

[41] Tesla. Tesla Vehicle Safety Report: Vehicle Fire Data. 2020. https://www.tesla.com/VehicleSafetyReport.

[42] Zhang G, Ge S, Xu T, Yang XG, Tian H, Wang CY. Rapid self-heating and internal temperature sensing of lithium-ion batteries at low temperatures. Electrochimica Acta. 2016;**218**:149-155

[43] Hu C. Enabling life After damage in lithium-ion batteries (NASA). In: 2020 NASA Aerospace Battery Workshop. 2020. https://www.nasa.gov/sites/default/files/atoms/files/nabw20_enab_life_after_damage_liion_battery_chu.pdf

[44] Corda S. In-Flight Vibration Environment of the NASA F-15B Flight Test Fixture. NASA Dryden Flight Research Center: National Technical Information Service (NTIS); 2002

[45] Muralidharan N. Mechano-Electrochemistry for Advanced Energy Storage and Harvesting Devices. Nashville, Tennessee, USA: Vanderbilt University; 2018

[46] Muralidharan N, Brock CN, Cohn AP, Schauben D, Carter RE, Oakes L, et al. Tunable mechanochemistry of lithium battery electrodes. ACS Nano. 2017;**11**(6):6243-6251

[47] Muralidharan N, Carter R, Oakes L, Cohn AP, Pint CL. Strain engineering to modify the electrochemistry of energy storage electrodes. Scientific Reports. 2016;**6**(1):1-9

[48] Kalnaus S, Wang Y, Turner JA. Mechanical behavior and failure mechanisms of Li-ion battery separators. Journal of Power Sources. 2017;**348**:255-263

[49] Hatzell KB, Chen XC, Cobb CL, Dasgupta NP, Dixit MB. Challenges in lithium metal anodes for solid state

batteries. ACS Energy Letters. 2020;**5**: 922-934

[50] Coelho ALV, Sandes NC. Data clustering via cooperative games: A novel approach and comparative study. Information Sciences. 2021;**545**:791-812

[51] Asano T, Sakai A, Ouchi S, Sakaida M, Miyazaki A, Hasegawa S. Solid halide electrolytes with high lithium-ion conductivity for application in 4 V class bulk-type all-solid-state batteries. Advanced Materials. 2018;**30**:1803075. DOI: 10.1002/adma.201803075

[52] Latroche M, Blanchard D, Cuevas F, El Kharbachi A, Hauback BC, Jensen TR, et al. Full cell hydride-based solid-state Li batteries for energy storage. International Journal of Hydrogen Energy. 2019;**44**:7875-7887

[53] Huen P, Ravnsbæk DB. All-solid-state lithium batteries – The Mg_2FeH_6-electrode $LiBH_4$-electrolyte system. Electrochemistry Communications. 2018;**87**:81-85. DOI: 10.1016/j.elecom.2018.01.001

[54] Houa W, Guoa X, Shena X, Amineb K, Yua H, Lub J. Solid electrolytes and interfaces in all-solid-state sodium batteries: progress and perspective. Journal of Power Sources. 2014;**247**:975-980. DOI: 10.1016/j.jpowsour.2013.09.051

[55] Zhou Q, Xu B, Chien P-H, Li Y, Huang B, Wu N, et al. NASICON $Li_{1.2}Mg_{0.1}Zr_{1.9}(PO_4)_3$ solid electrolyte for an all-solid-state Li-metal battery. Small Methods. 2020;**9**:1-7. DOI: 10.1002/smtd.202000764

[56] Parejiya A, Essehli R, Amin R, Liu J, Muralidharan N, Meyer HM III, et al. Na1+xMnx/2Zr2−x/2(PO4)3 as a Li+ and Na+ super ion conductor for solid-state batteries. ACS Energy Letters. 2021; **6**:429. DOI: 10.1021/acsenergylett.0c02513

[57] Viswanathan et al. The challenges and opportunities of battery-powered flight. Nature. 2022;**601**:519-525

[58] Yang XG, Liu T, Gao Y, Ge S, Leng Y, Wang D, et al. Asymmetric temperature modulation for extreme fast charging of lithium-ion batteries. Joule. 2019;**3**:3002-3019

[59] Ge S, Leng Y, Liu T, Longchamps RS, Yang XG, Gao Y, et al. A new approach to both high safety and high performance of lithium-ion batteries. Science Advances. 2020;**6**:eaay7633

Chapter 5

Construction and Modification of Copper Current Collectors for Improved Li Metal Batteries

Shunrui Luo and Kai Pei

Abstract

Metallic Lithium have gained great attention for its high theoretical specific capacity. But continuous growth of Li dendrites upon cycling might cause low coulombic efficiency and serious security issues. Construction of advanced 3D Cu current collectors to regulate Li plating/stripping and improve battery performance is considered as one effective promising strategy. In this chapter, we will discuss the roles and requirements of current collectors in lithium metal batteries. Then methods (dealloying, powder-sintering and 3D printing) employed for construction of 3D Cu current collector and implementation of surface modification (lithiophilic sites and coating layers) will be illustrated. At last, future opportunities of Cu current collectors will be lifted out.

Keywords: Cu current collectors, Li deposition, 3D framework, surface modification, Lithium metal batteries

1. Introduction

With the increasing demand for electric vehicles and portable electronics, there is an urgent need for advanced energy storage systems with high energy density [1]. State-of-the-art lithium-ion batteries (LIBs) with intercalated graphite as anodes and transition metal oxide as cathodes are currently approaching their theoretically energy densities (~300 Wh/kg), which is still insufficient to meet the growing demand of advanced energy storage system and achieve a carbon neutral society [2]. Li metal have gained much attention as anode for its high theoretical specific capacity (3860 mAh g^{-1}), low redox potential (−3.04 V vs. SHE), and low density of 0.59 g cm^{-3}, which could boost the energy density of batteries [3]. However, lithium anode has many notorious problems during the cycling process, namely: formation of dendrites, corrosion of lithium, dead lithium, and volume expansion, which will lead to several capacity loss and even explosion of lithium metal batteries [3, 4]. Specifically speaking, the fragile SEI (solid electrolyte interphase) would crack due to huge volume expansion during Li plating (**Figure 1**, step 1), through which Li dendrite will grow (inhomogeneous Li deposition; step 2) [5]. During Li stripping, SEI would be also cracked due to the volume shrink, and 'dead Li' (electrically isolated from the substrate) will form from kinks or roots in a

IntechOpen

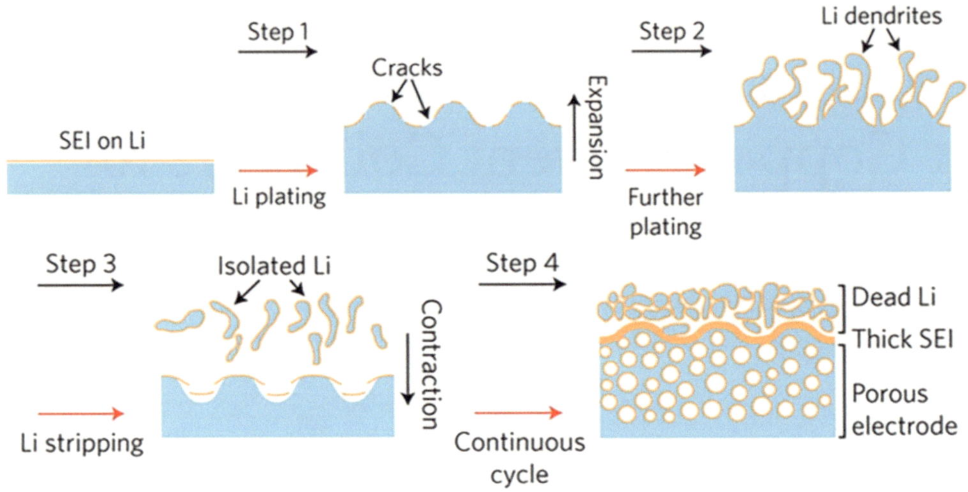

Figure 1.
Correlations among the different challenges in the Li metal anode, originating from high reactivity and infinite relative volume change (reproduced from ref. [5]).

dendrite (step 3). The repeated process will amplify the problem, and finally lead to a porous Li electrode, a thick accumulated SEI layer and excessive dead Li (step 4). The nucleation and growth of lithium dendrites are mainly influenced by types of nucleation [6, 7], distribution of Li ions [8, 9], and charge induction [10]. Until now, various suppression strategies for Li dendrites have been reported, including modification of electrolyte [11, 12], application of different temperatures and pressures [13, 14], fabrication of multifunctional separators [14, 15], etc. Among them, building advanced current collectors are considered an outstanding technique for the suppression of dendritic growth and stable Li anodes in safe Li metal batteries (LMBs) [16, 17].

Current collector is one of the essential components of LMBs, which not only supports anode or cathode materials layers, but also collects electrons from electrochemical reaction, and bridges with external circuits. The commercially used cathode and anode current collectors for Lithium-ion battery are Aluminum (Al) and copper (Cu) foil respectively. Researched results showed that current collectors have great influence on the resistance and performance of lithium metal batteries [18]. In other words, rate capacity, coulombic efficiency and cycling capacity of LMBs would be enhanced by increasing conductivity, lowering contact resistance, and improving chemical resistance of current collectors. In this chapter, we will introduce the general requirements and challenges for Cu collectors, and strategies employed for construction and modification of copper current collectors for improved Li metal batteries anodes.

2. Copper current collectors

The ideal current collectors for LIBs should involve the following four conditions. (1) High chemical resistance. Current collectors should be chemically stable to active materials and electrolyte during manufacturing and charge/discharge processes. (2) High electrical conductivity. High electrical conductivity of current collectors allows efficient collecting and transporting of electrons, improved battery performances and alleviated conversion of chemical/electrical energy to heat which would rise the working temperature of LIBs. (3) Lightweight and flexible. Elevated volumetric and gravimetric energy densities could be achieved with lightweight and ultrathin current collectors. Flexible current collectors with elevated mechanical strength could resist the strain and keep their integrity during manufacturing and cycling process. (4) low-cost and processability. Low-cost and processability of current collectors are very vital for its commercialization.

Copper foil (with a thickness of 5–12 μm) is widely applied as anode current collectors for lithium metal batteries because of its low resistivity (~1.68 * 10^{-8} Ω m, 20°C), good electrochemical stability from 0 to 3 V vs. Li/Li^+ in $LiPF_6$ electrolyte, low tensile strength (~ 325 Mpa), affordable price ($ 640/m^2) and easy access [17]. Until now, Cu current collector structures of Cu foil, Cu mesh and 3D Cu frame works (including Cu foam, coated Cu, and other novel 3D Cu structures) have been reported [17]. The lithiophobicity of Cu usually causes the inhomogeneous distribution of Lithium ions (Li^+) and the formation and growth of small Lithium dendrites. The small Li dendrites lead to inhomogeneous distribution of the electric field and Li^+ concentration, which finally promote, under cycling, the growth of dangerous dendritic and mossy Li (**Figure 2a**) [19]. Li dendrites could penetrate the separator and result in serious capacity decay by a consumption of both Li and electrolyte. This hinders the application of Cu foil and Cu mesh as current collectors. 3D Cu framework with large internal surface area and room could accommodate Li deposition, lowering the electrode current density to suppress/reduce Li dendrite growth and alleviate the huge volume change of Li metal during cycling (**Figure 2b**) [19]. The presence of lithiophilic sites on the surface of current collectors could decrease the overpotential and result in homogeneous lithium deposition.

In the next section (section3), we will provide a mini-review of some of the methods used for the design and construction of 3D Cu framework as current collectors for Lithium metal batteries. In the Section 4, we will present some of the strategies employed to modify the surface of Cu current collectors and tailor its lithiophobicity.

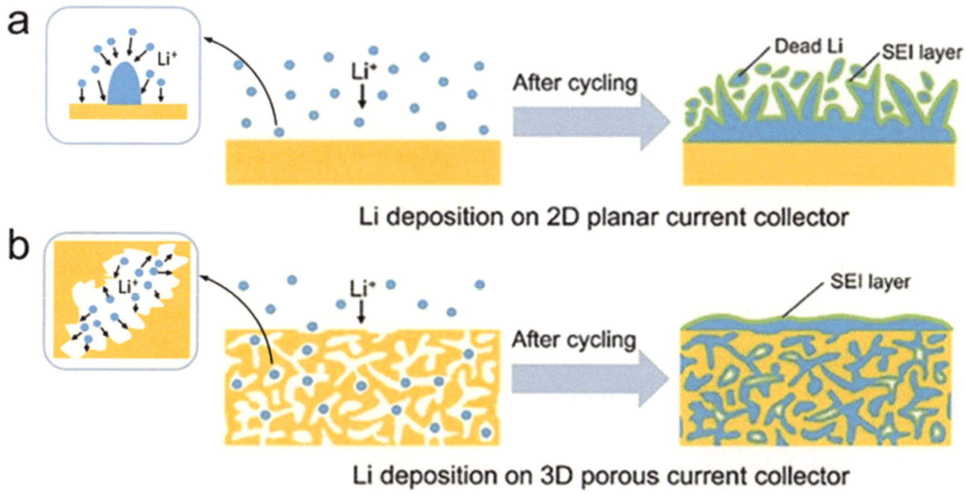

Figure 2.
A schematic view of the structural changes during Li plating/striping in a conventional 2D planar current collector (a) and a 3D porous current collector (b) (reproduced from ref. [19]).

3. 3D copper frameworks

Alloying/Dealloying is a widely used method for the manufacturing of porous 3D copper current collector by selective "splitting" of metals based on the different chemical, electrochemical and physical properties of the components. Dealloying bimetallic Cu-Zn alloy (brass) [19] in a mixture solution of HCl and NH_4Cl resulted in dissolution of zinc (Zn) and formation of 3D interconnected copper framework. When used as current collectors, the porous Cu skeleton facilitated electron transport and the interconnected pores helped to accommodate the Li deposition. This 3D copper current collector showed excellent performance (97% of Coulombic efficiency for 250 cycles at 0.5 mA cm^{-2}) and has commercial potentials in Li-ion batteries. The follow-up works of Yan-bing He *et al.* [16] showed that a more precisely engraved 3D porous structure could be obtained with electrochemical method by controlling the cycles of electrochemical etching process. The obtained 3D Cu structure with uniform and smooth pores also showed superior mechanical properties and good electrical conductivity. The lithium anode with such 3D Cu current collector achieved stable cycling for 400 h at a high Li plating/stripping capacity of 1 mA h cm^{-2} and high current density of 1 mA cm^{-2}.

Inspired by the fact that lithiophilic metals such as Au, Ag, Mg and Zn presented low nucleation potentials (essentially 0 V) of lithium on them and could promote uniform Li deposition, Jun Lu *et al.* [20] put forward Cu-Zn alloys (commercially available brass) as lithiophilic materials based on DFT calculations and *in situ* experiments. Lithiophilic 3D porous Cu-Zn current collector with porous structure and residual CuZn alloy has been prepared by facile chemical dealloying Cu-Zn sheets in hydrochloric acid at various dealloying time (from 0 to 8 h, **Figure 3a–h**). Li dendrites were more suppressed in 3D porous Cu (8 h-3D Cu) than planar Cu which is consistent with what we discussed above. The 2 h-3D CuZn current collector showed an optimal lithiophilicity with denser and more uniform Li layers (**Figure 3i–l**), and a high CE (>95% of coulombic efficiency after 220 cycles at 0.5 mA cm^{-2}) over 4 h Cu-Zn (97.5% of coulombic efficiency after 135 cycles at 0.5 mA cm^{-2}) for elevated Li preferred nucleation sites in the 2 h-3D framework. A painting-alloying-dealloying strategy [21] was developed to fabricate 3D porous Cu by painting melted Ga on Cu foil, alloying at 100°C and chemically dealloying in a solution of nitric acid and hydrofluoric acid. There are also reports on chemical dealloying of alloys like Cu-Al-Mn [22], Al-Cu [23] in solution of HCl and NaOH respectively to prepare 3D porous Cu current collectors. Apart from chemical dealloying method, Cu-Zn alloy can also be dealloyed utilizing a vacuum distillation approach (physical vacuum dealloy) owing to the easily distillated Zn component (bp, 907°C) [24]. As for current collectors based on free-standing copper nanowires network [25], their open porous pores enabled the Li metal to be deposited on the whole surface of the network and fill the pores without dendritic lithium growth.

Powder-sintering is a heat treatment applied to compact powder in order to improve the strength and structure integrity. It is, thought to be a facile, effective, and low-cost way for the preparation of porous materials [26, 27]. The particles of the powder chemically bond to each other and form a consistent shape under a high temperature (below the melting point of the main component). For example, an integrated bidirectional porous Cu (IBP-Cu) film was obtained by powder sintering of commercial Cu powder at a certain temperature range (from 700 to 900°C) [27]. This IBP-Cu film with tunable porosity presented a strong mechanical flexibility and stability. Li/IBP-Cu anode could achieve a capacity of up to 7.0 mAh cm^{-2} for deep plating/stripping and long

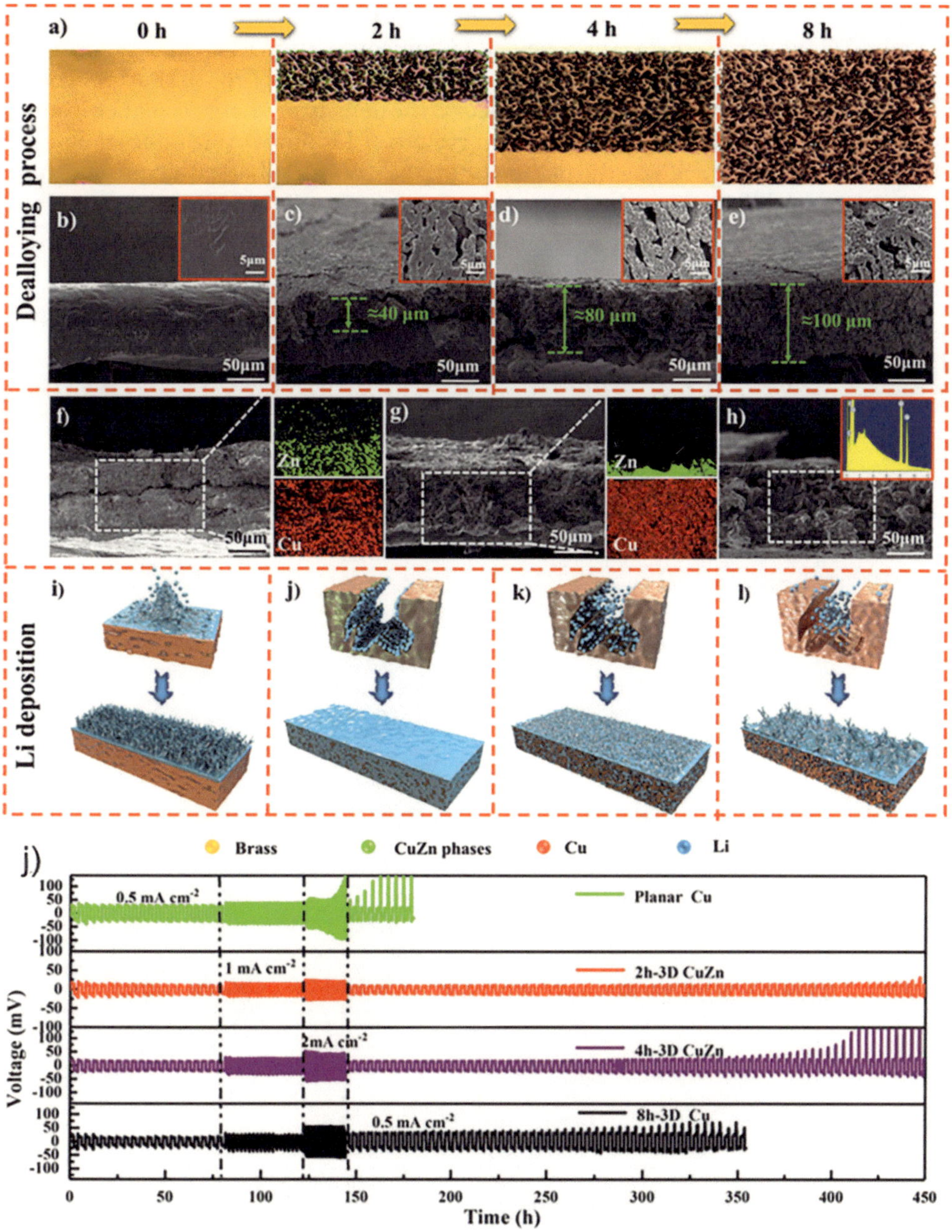

Figure 3.
(a) Dealloying process schematic diagram (CuZn phases were dealloyed to Cu phases at different time duration). (b–e) SEM cross-sectional and upper surface images of CuZn sheets before and after dealloying at 0, 2, 4, and 8 h. (f–h) EDX cross-sectional images after dealloying at 2, 4, and 8 h. (i–l) Li deposition diagram indicating displaying different current collectors dealloyed with various durations (at 0, 2, 4, and 8 h). (j) Rate performance of different current collectors as Li metal anodes (reproduced from ref. [20]).

cycling ability with Coulombic efficiency of ≈100% for 1000 cycles at 1.0 mA cm^{-2} and 1.0 mAh cm^{-2}. NaCl-assisted powder-sintering (using NaCl as submicron templates) allowed to improve porosity of Cu current collectors and prevent congestion of

Li-diffusion during Li plating process [28]. As already reported above, 3D Cu skeleton structure [28] with high surface area helped Li homogeneous plating: Li is deposited largely in the internal pores of 3D skeleton rather than outside of the 3D current collectors. Indeed, following the lightning rod theory, the density of charge is higher in areas of high curvature leading to the preferential growth of Lithium inside internal pores [29]. Smooth surfaces facilitate the plating of Li into smooth spherical shape, which

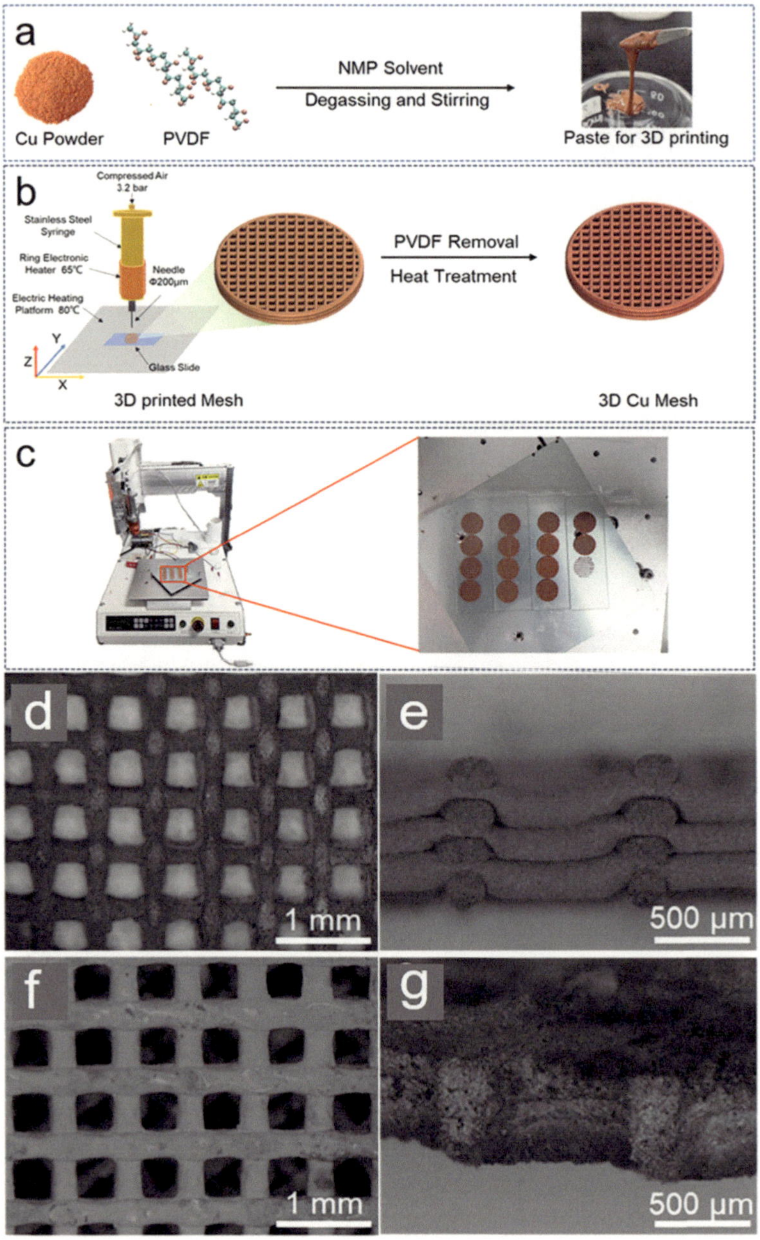

Figure 4.
Schematic illustration of the formation of a 3D printed Cu mesh. (a) Preparation of paste. (b) Schematic diagram of the 3D printing mechanism and postprocessing process of the Cu mesh. (c) Mass-produced 3D printed Cu meshes. Morphology of the 3D Cu mesh before (d–e) and after (f–g) heat treatment (reproduced from ref. [37]).

ease Li dendrite growth. As a result, the 3D current collectors based on NaCl assisted powder-sintering [28] performed a high and stable Coulombic efficiency (higher than 95% for 700 cycles at 0.5 mA cm^{-2} and 400 cycles at 1 mA cm^{-2} respectively). Moreover, introduction of lithiophilic components like Zn before powder-sintering would enhance the lithiophilicity of the whole current collector [27, 30].

Additive manufacturing (AM, also known as 3D printing) [31] has been employed for the facile and accurate fabrication of 3D architectures by a layer-by-layer depositing phase change materials and solvent-based inks from computer-aided-design (CAD) files. This method has attracted high interests in various areas not limited to the design of 3D architectures as components for energy storage [32–36]. Various 3D print techniques, such as DIW (direct ink writing), IJP (inkjet printing), SLA (stereolithography), FDM (fused deposition modeling), *etc.*, have been employed to manufacture micro-sized components for lithium batteries [31]. For example, a hierarchical 3D Cu current collector [37] was obtained *via* DIW of copper ink followed with pyrolysis and reduction (**Figure 4a**). The printing paste was prepared by mixing Cu powder, polyvinylidene fluoride (PVDF), and 1-methyl-2-pyrrolidinone (NMP). High viscosity of the paste can be realized by minimizing the volume of the added NMP, which helps to avoid the collapsing of the extruded lines. A ring electric heater was equipped on the syringe to improve the fluidity of the paste for better extrusion (**Figure 4b**). The extruded paste was heated on a hot plate to vaporize NMP and induce solidification (**Figure 4c**). The 3D Cu mesh configuration [37] was finally obtained by heat treatment in air to remove PVDF and then in hydrogen for the reduction of CuO completely to Cu. The large pores of 3D Cu current collector help to store electrolyte and provide spaces for dendrite growth. The small pores generated from the spaces occupied by PVDF (polyvinylidene fluoride) and the space where particles accumulated during heat treatment, would guide dendrite deposition. Thus, the Cu current collector-based anode showed enhanced deposition and stripping capability (50 mAh cm^{-2}), high-rate capability (50 mA cm^{-2}), and a long-term stable cycling (1000 h). The 3D printing ink formula and printing process can be guided by machine learning [38]. In another example, a 3D-printed Cu (3DP-Cu) was fabricated by extrusion DIW approach with a mixture of copper particles, Pluronic F127 (PF127) and water used as printing ink [39]. The 3DP-Cu exhibited good mechanical stability and robustness even under a high pressure of around 1.4 MPa.

4. Surface modification

Apart from constructing 3D current collectors to attain uniform striping and plating of lithium, smart surface modification of copper-based current collectors to enhance lithiophilicity and lower interfacial energy between lithium and the substrate could also favor uniform deposition of Lithium. Next, we will mainly discuss some of the strategies employed for the surface modification of Cu-based current collectors.

4.1 Introducing lithiophilic sites

Lithiophobicity of Cu constrains its application as current collector for Li Metal based batteries due to high deposition overpotential of lithium. Introduction of lithiophilic sites could decrease the overpotential and result in smooth lithium deposition. Lithium nucleation overpotential of a list of 11 elements (Au, Ag, Zn, Mg, Al, Pt, Si, Sn, C, Cu, and Ni) substrate materials were carefully studied by vacuum deposition of these materials (with a thickness of 50 nm) on Cu foil as working electrode. The related voltage profiles are shown in **Figure 5(a** and **b)** [40]. Voltage profile of Cu shows a major voltage dip and a followed flat voltage plateau (10 μA cm^{-2}) (**Figure 5(b)** red line), while Au, Ag, Zn and Mg showed zero overpotential during deposition of Li. When Cu substrate patterned with Au stripes was used for galvanostatic deposition, Li was found on the surface of Au strips, whereas no Li were found on the bare Cu surface (**Figure 5c**). C, Sn and Si showed a lower Li nucleation overpotential compared with Ni and Cu. In addition, the element Sb [41], other metal compounds (like Cu_2O [42], CoO [43], etc.) and polymers (like polydopamine, etc.) were also shown to have a low overpotential and trigger an uniform Li deposition when coated onto Cu current collectors. These materials are usually applied as a coating layer on the current collectors and will be discussed in the next section.

4.2 Create a coating layer

The coating of Cu current collectors with various materials offers substantial benefits. Carbonaceous materials have been widely used for the fabrication of core-shell structured composites and also applied in energy storage for their high conductivity, lightweight, and good chemical stability, etc. 3D Cu foams wrapped with N-doped porous carbon nanosheets (with a thickness from several to 35 nm, NPCN) were obtained by dip-coating process [44]. The lithiophilicity and large surface area of such NPCN could help to regulate Li deposition and avoid Li dendrites even at a high Li deposition capacity of 10 mAh cm^{-2}. The Li/Cu@NPCN||Li cell demonstrated a high coulombic efficiency and a low hysteresis after cycling for 1600 h (0.5 mA cm^{-2}). Graphene film modified Cu foam doped with Br- and CuBr- (BGCF) could guide Li^+ flux and therefore achieve an homogeneous Li deposition [45]. Also, CuBr are able to react with Li to form LiBr which offers a fast Li diffusion pathway. Pancake-like Li seeds and further Li metal were observed under a current density of 2 mA cm^{-2}. Li||Li-BGCF symmetrical cell attained a cycling stability of 850 h (1 mA cm^{-2}, **Figure 6c**). Li||$LiFePO_4$ full batteries fabricated with those BGCF maintained 94% of their initial capacity after 400 cycles with a cathode loading of 2 mAh cm^{-2}. Besides, porous carbon nanosheets, carbon-based materials like graphene [47], carbon nanofibers [48] wrapped Cu current collectors have been reported to regulate Li deposition and alleviate Li dendrites. The addition of carbon-based materials improves chemical stability and charge transfer kinetics. Note that the thickness of carbon coating layer should be precisely controlled and parameters like porosity, surface treatment, etc. should be investigated for a further understanding of the mechanism.

The growth of vertically aligned CuO nanosheets on planar Cu (VA-CuO NSs) were realized by immersing Cu foil in NH_4OH solution [42]. The intimate contact between CuO nanosheets and Cu foil ensures the good conductivity of the electrode. The CuO nanosheets enabled increased affinity with Li, which ensures uniform Li nucleation. When exploited as current collectors, the Li|VA-CuO-Cu preserved 94% of the coulombic efficiency after 180 cycles (1 mA cm^{-2}). A Cu foils with Cu pillar arrays were fabricated by templated electrodeposition with controllable pillars diameters and pore spacing (**Figure 6d** and **e**) [46]. This vertical pillar architecture could block Li^+ flux on the top of pillars and result in Li deposited within the pores rather than on the top of the pillars arrays, which reduced safety issues. When coated with a thin layer (50 nm) of ZnO by atomic layer deposition (ALD), this core-shell pillar architecture facilitated uniform Li deposition. With synergistic effect of ALD surface modification, the cycling lifetime were tripled (>180 h) [46]. Until now, Sb

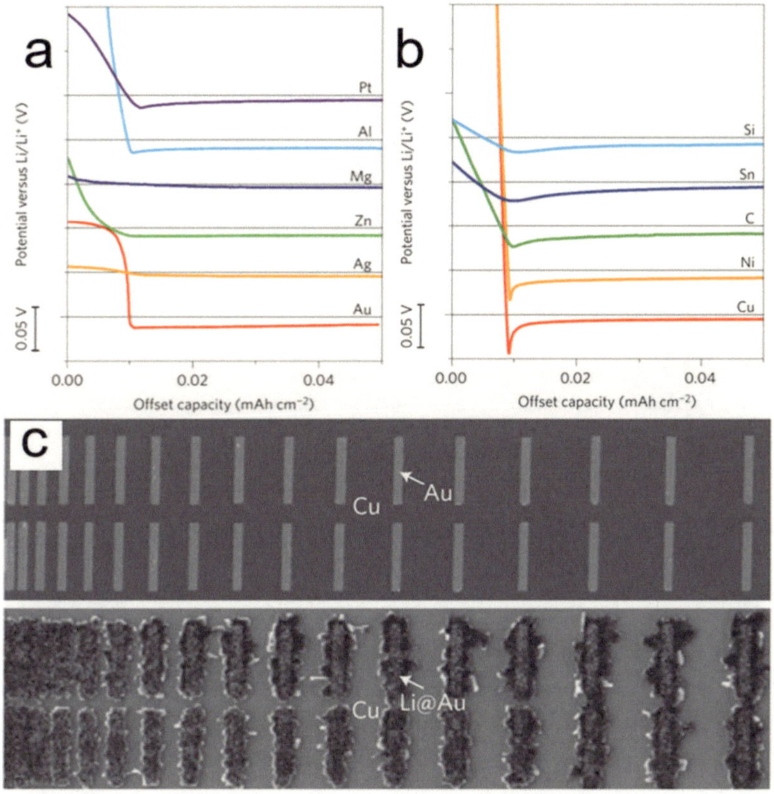

Figure 5.
Voltage profiles of various materials with certain solubility (a) and negligible (b) solubility in Li during Li deposition at a current density of 10 μA cm^{-2}. The horizontal grey lines display 0 V versus Li/Li^+ in (a) and (b). (c) SEM images of a gold strip array with various separations (increased from 1 μm (left) to 16 μm (right)) before (top) and after (bottom) Li deposition (reproduced from ref. [40]).

[41], CoO [43], etc. have been reported to increase lithiophilicity and hence regulate the Li nucleation. With the addition of a coating layer, the thickness and conductivity of the whole current collector should be carefully balanced.

When a thin β-PVDF (Poly(vinylidene difluoride)) layer with high dielectric constant (8–13) was coated on copper foil, the formed β-PVDF@Cu attained dendrite-free Li deposition up to 5 mA cm^{-2} and enhanced battery performance for Li metal anodes [49]. Firstly, the polarity of PVDF redistributed Li-ion flux through the interactions between Li ions and its polar groups. Secondly, the alignment of F atoms facilitated uniform Li deposition for interaction between the electronegative C-F groups and Li at the interface. Also, F atoms on the PVDF induce preferential Li diffusion path and allow to achieve high-rate Li plating/stripping. For polydopamine (PDA) modified Cu foil [50], PDA reacts with Li *via* the carbonyl groups complexing with Li. This uniform distribution of Li ions helps to achieve uniform Li deposition on the surface of current collector. In addition, conductive polymers like poly(dimethylsiloxane) (PDMS) [51], poly(3,4-ethylenedioxythiophene (PEDOT) [52], polyaniline (PANi) [52], etc. have been reported for the modification of Cu foil as improved current collectors. Conductive polymers are potential candidates for current collectors thanks to their low densities, good mechanical strength, and chemical resistance. Nevertheless, the conductivity and thermal stability need be improved to meet commercial requirements.

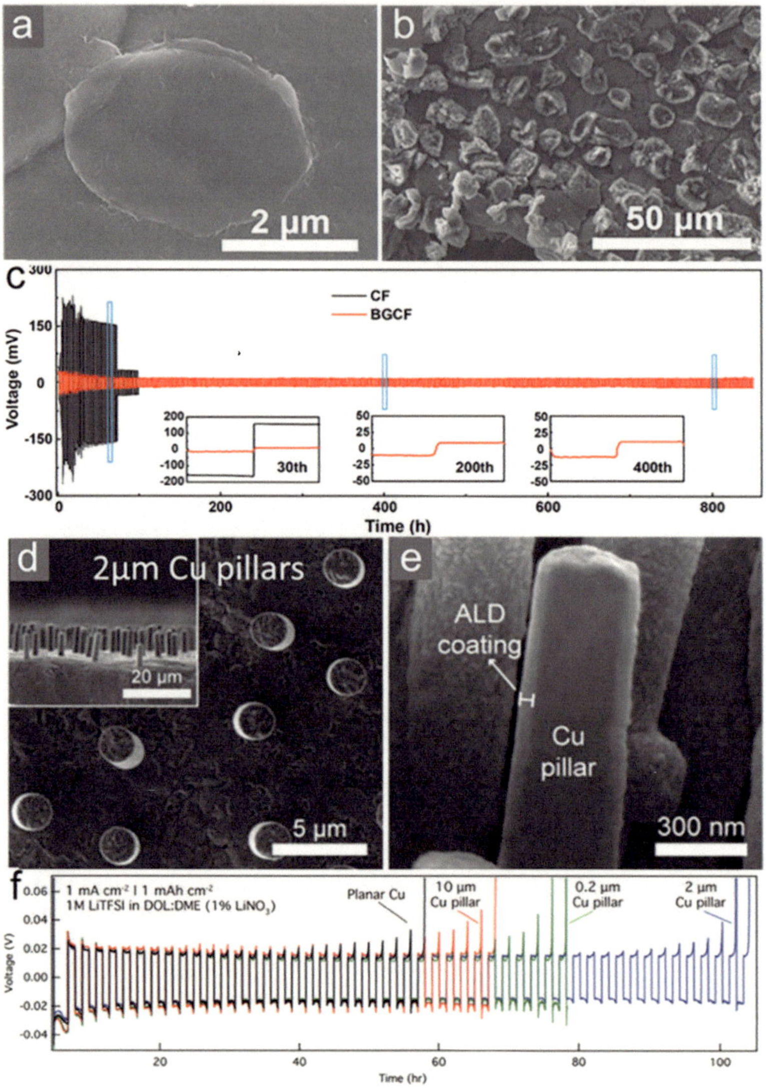

Figure 6.
SEM images after plating of Lithium metal at 0.5 mAh cm^{-2} (a) and stripping at 2 mAh cm^{-2} (b) on BGCF. (c) Voltage–time curves of Li–CF and Li–BGCF in a symmetric cell at 1 mA cm^{-2} for 1 mA h cm^{-2} [45]. SEM images of Cu pillars (d) and core-shell ALD-coated Cu pillar structure (e). (f) Galvanostatic cycling of Li–Cu cells at 1 mA cm^{-2} for 1 mAh cm^{-2} on the planar Cu and Cu pillars with various diameters (0.2 μm, 2 μm, and 10 μm, reproduced from ref. [46]).

5. Conclusion

The Li metal battery technology has attracted great attentions for their high energy capacities and low density. However, poor chemical stability, growth of Li dendrites, and unstable electrode-electrolyte interfaces hinder its utilization. One of the effective strategies is the development of advanced current collectors to regulate Li plating/stripping and improve performances. 3D Copper current collectors with a large surface area could lower local current density and constrain volume changes, which would effectively prevent lithium dendrites growth and mitigate volume changes. The methods of dealloying, powder-sintering and 3D printing have been discussed for the realization of 3D Cu frameworks. Surface modification strategies mainly includes introduction of lithiophilic sites through coating layers (metal composites, carbon-based materials and polymers) to regulate Li ion deposition and avoid Li dendrites. Thus, stable coulombic efficiencies and prolonged cycling lives were achieved. As shown in this chapter, the development of advanced Cu current collectors to fabricate high gravimetric/volumetric energy densities Li metal batteries with long cycling life will have to consider all the parameters, like porosities, thickness, morphologies, cost, feasibility, etc., and utilize not only one strategy for their construction. For future current collectors, the mechanism governing Li dendritic growth will also need more in-depth investigation, more precise preparation control to promote their actual application and achieve mass production.

Acknowledgements

This work was financially supported by the science and technology innovation Program of Hunan Province (2021RC2007) and China Postdoctoral Science Foundation project (2020 M682700).

Author details

Shunrui Luo[1] and Kai Pei[2]*

1 School of Metallurgy and Environment, Engineering Research Center of the Ministry of Education for Advanced Battery Materials, Central South University, Changsha, China

2 School of Environment and Energy, South China University of Technology, Guangzhou, China

*Address all correspondence to: peikai@scut.edu.cn

References

[1] Fichtner M, Edström K, Ayerbe E, Berecibar M, Bhowmik A, Castelli IE, et al. Rechargeable batteries of the future—The state of the art from a BATTERY 2030+ perspective. Advanced Energy Materials:2102904

[2] Deng D. Li-ion batteries: Basics, progress, and challenges. Energy Science & Engineering. 2015;**3**:385-418

[3] Liu D-H, Bai Z, Li M, Yu A, Luo D, Liu W, et al. Developing high safety Li-metal anodes for future high-energy Li-metal batteries: Strategies and perspectives. Chemical Society Reviews. 2020;**49**:5407-5445

[4] Horstmann B, Shi J, Amine R, Werres M, He X, Jia H, et al. Strategies towards enabling lithium metal in batteries: Interphases and electrodes. Energy & Environmental Science. 2021;**14**:5289-5314

[5] Lin D, Liu Y, Cui Y. Reviving the lithium metal anode for high-energy batteries. Nature Nanotechnology. 2017;**12**:194-206

[6] Ely DR, García RE. Heterogeneous nucleation and growth of Lithium electrodeposits on negative electrodes. Journal of the Electrochemical Society. 2013;**160**:A662-A668

[7] Harry KJ, Hallinan DT, Parkinson DY, MacDowell AA, Balsara NP. Detection of subsurface structures underneath dendrites formed on cycled lithium metal electrodes. Nature Materials. 2014;**13**:69-73

[8] Hao F, Verma A, Mukherjee PP. Mechanistic insight into dendrite–SEI interactions for lithium metal electrodes. Journal of Materials Chemistry A. 2018;**6**:19664-19671

[9] Yamaki J-I, Tobishima S-I, Hayashi K, Keiichi S, Nemoto Y, Arakawa M. A consideration of the morphology of electrochemically deposited lithium in an organic electrolyte. Journal of Power Sources. 1998;**74**:219-227

[10] Chazalviel JN. Electrochemical aspects of the generation of ramified metallic electrodeposits. Physical Review A. 1990;**42**:7355-7367

[11] Basile A, Bhatt AI, O'Mullane AP. Stabilizing lithium metal using ionic liquids for long-lived batteries. Nature Communications. 2016;**7**:ncomms11794

[12] Zhao C-Z, Zhao B-C, Yan C, Zhang X-Q, Huang J-Q, Mo Y, et al. Liquid phase therapy to solid electrolyte–electrode interface in solid-state Li metal batteries: A review. Energy Storage Materials. 2020;**24**:75-84

[13] Geng Z, Lu J, Li Q, Qiu J, Wang Y, Peng J, et al. Lithium metal batteries capable of stable operation at elevated temperature. Energy Storage Materials. 2019;**23**:646-652

[14] Wu F, Yuan Y-X, Cheng X-B, Bai Y, Li Y, Wu C, et al. Perspectives for restraining harsh lithium dendrite growth: Towards robust lithium metal anodes. Energy Storage Materials. 2018;**15**:148-170

[15] Liu H, Wang E, Zhang Q, Ren Y, Guo X, Wang L, et al. Unique 3D nanoporous/macroporous structure Cu current collector for dendrite-free lithium deposition. Energy Storage Materials. 2019;**17**:253-259

[16] Zhao H, Lei D, He Y-B, Yuan Y, Yun Q, Ni B, et al. Compact 3D copper with uniform porous structure derived by electrochemical Dealloying as dendrite-free Lithium metal anode current collector. Advanced Energy Materials. 2018;**8**:1800266

[17] Zhu P, Gastol D, Marshall J, Sommerville R, Goodship V, Kendrick E. A review of current collectors for lithium-ion batteries. Journal of Power Sources. 2021;**485**:229321

[18] Yamada M, Watanabe T, Gunji T, Wu J, Matsumoto F. Review of the Design of Current Collectors for improving the battery performance in Lithium-ion and post-Lithium-ion batteries. Electrochem. 2020;**1**:124-159

[19] Yun Q, He Y-B, Lv W, Zhao Y, Li B, Kang F, et al. Chemical Dealloying derived 3D porous current collector for Li metal anodes. Advanced Materials. 2016;**28**:6932-6939

[20] Zhang D, Dai A, Wu M, Shen K, Xiao T, Hou G, et al. Lithiophilic 3D porous CuZn current collector for stable Lithium metal batteries. ACS Energy Letters. 2020;**5**:180-186

[21] Shi Y, Wang Z, Gao H, Niu J, Ma W, Qin J, et al. A self-supported, three-dimensional porous copper film as a current collector for advanced lithium metal batteries. Journal of Materials Chemistry A. 2019;**7**:1092-1098

[22] Luo Z, Xu JC, Yuan B, Li H, Hu RZ, Yang LC, et al. A novel 3D bimodal porous current collector with large interconnected spherical channels for improved capacity and cycling stability of Sn anode in Li-ion batteries. Materials Letters. 2018;**213**:189-192

[23] Zhang S, Xing Y, Jiang T, Du Z, Li F, He L, et al. A three-dimensional tin-coated nanoporous copper for lithium-ion battery anodes. Journal of Power Sources. 2011;**196**:6915-6919

[24] An Y, Fei H, Zeng G, Xu X, Ci L, Xi B, et al. Vacuum distillation derived 3D porous current collector for stable lithium–metal batteries. Nano Energy. 2018;**47**:503-511

[25] Lu L-L, Ge J, Yang J-N, Chen S-M, Yao H-B, Zhou F, et al. Free-standing copper nanowire network current collector for improving Lithium anode performance. Nano Letters. 2016;**16**:4431-4437

[26] Chen J, Li S, Qiao X, Wang Y, Lei L, Lyu Z, et al. Integrated porous Cu host induced high-stable bidirectional Li plating/stripping behavior for practical Li metal batteries. Small. 2022;**18**:2105999

[27] Fan H, Dong Q, Gao C, Hong B, Lai Y. Powder-sintering derived 3D porous current collector for stable lithium metal anode. Materials Letters. 2019;**234**:69-73

[28] Wang Y, Wang Z, Lei D, Lv W, Zhao Q, Ni B, et al. Spherical Li deposited inside 3D Cu skeleton as anode with Ultrastable performance. ACS Applied Materials & Interfaces. 2018;**10**:20244-20249

[29] Park J, Jeong J, Lee Y, Oh M, Ryou M-H, Lee YM. Micro-patterned Lithium metal anodes with suppressed dendrite formation for post Lithium-ion batteries. Advanced Materials Interfaces. 2016;**3**:1600140

[30] Qiang Z, Liu B, Yang B, Ma P, Yan X, Wang B. Preparation of three-dimensional copper-zinc alloy current

collector by powder metallurgy for Lithium metal battery anode. ChemElectroChem. 2021;**8**:2479-2487

[31] Pei M, Shi H, Yao F, Liang S, Xu Z, Pei X, et al. 3D printing of advanced lithium batteries: A designing strategy of electrode/electrolyte architectures. Journal of Materials Chemistry A. 2021;**9**:25237-25257

[32] Liu F, Chen Q, Liu C, Ao Q, Tian X, Fan J, et al. Natural polymers for organ 3D bioprinting. Polymers. 2018;**10**:1278

[33] Godoi FC, Prakash S, Bhandari BR. 3d printing technologies applied for food design: Status and prospects. Journal of Food Engineering. 2016;**179**:44-54

[34] Muth JT, Vogt DM, Truby RL, Mengüç Y, Kolesky DB, Wood RJ, et al. Embedded 3D printing of strain sensors within highly stretchable elastomers. Advanced Materials. 2014;**26**:6307-6312

[35] Browne MP, Redondo E, Pumera M. 3D printing for electrochemical energy applications. Chemical Reviews. 2020;**120**:2783-2810

[36] Chang P, Mei H, Zhou S, Dassios KG, Cheng L. 3D printed electrochemical energy storage devices. Journal of Materials Chemistry A. 2019;**7**:4230-4258

[37] Chen C, Li S, Notten PHL, Zhang Y, Hao Q, Zhang X, et al. 3D printed Lithium-metal full batteries based on a high-performance three-dimensional anode current collector. ACS Applied Materials & Interfaces. 2021;**13**:24785-24794

[38] Cipollone D, Yang H, Yang F, Bright J, Liu B, Winch N, et al. 3D printing of an anode scaffold for lithium batteries guided by mixture design-based sequential learning. Journal of Materials Processing Technology. 2021;**295**:117159

[39] Lim GJH, Lyu Z, Zhang X, Koh JJ, Zhang Y, He C, et al. Robust pure copper framework by extrusion 3D printing for advanced lithium metal anodes. Journal of Materials Chemistry A. 2020;**8**:9058-9067

[40] Yan K, Lu Z, Lee H-W, Xiong F, Hsu P-C, Li Y, et al. Selective deposition and stable encapsulation of lithium through heterogeneous seeded growth. Nature Energy. 2016;**1**:16010

[41] Wang J-R, Wang M-M, He X-D, Wang S, Dong J-M, Chen F, et al. A Lithiophilic 3D conductive skeleton for high performance Li metal battery. ACS Applied Energy Materials. 2020;**3**:7265-7271

[42] Zhang C, Lv W, Zhou G, Huang Z, Zhang Y, Lyu R, et al. Vertically aligned Lithiophilic CuO Nanosheets on a Cu collector to stabilize Lithium deposition for Lithium metal batteries. Advanced Energy Materials. 2018;**8**:1703404

[43] Zhang R, Li Y, Wang M, Li D, Zhou J, Xie L, et al. Super-assembled hierarchical CoO Nanosheets-Cu foam composites as multi-level hosts for high-performance Lithium metal anodes. Small. 2021;**17**:2101301

[44] Pei F, Fu A, Ye W, Peng J, Fang X, Wang M-S, et al. Robust Lithium metal anodes realized by Lithiophilic 3D porous current collectors for constructing high-energy Lithium–Sulfur batteries. ACS Nano. 2019;**13**:8337-8346

[45] Duan H, Zhang J, Chen X, Zhang X-D, Li J-Y, Huang L-B, et al. Uniform nucleation of Lithium in 3D current collectors via bromide intermediates for stable cycling Lithium metal batteries. Journal of the American Chemical Society. 2018;**140**:18051-18057

[46] Chen K-H, Sanchez AJ, Kazyak E, Davis AL, Dasgupta NP. Synergistic

effect of 3D current collectors and ALD surface modification for high coulombic efficiency Lithium metal anodes. Advanced Energy Materials. 2019;**9**:1802534

[47] Yang G, Chen J, Xiao P, Agboola PO, Shakir I, Xu Y. Graphene anchored on Cu foam as a lithiophilic 3D current collector for a stable and dendrite-free lithium metal anode. Journal of Materials Chemistry A. 2018;**6**:9899-9905

[48] Chen Y, Elangovan A, Zeng D, Zhang Y, Ke H, Li J, et al. Vertically aligned carbon nanofibers on Cu foil as a 3D current collector for reversible Li plating/stripping toward high-performance Li–S batteries. Advanced Functional Materials. 2020;**30**:1906444

[49] Luo J, Fang C-C, Wu N-L. High polarity poly(vinylidene difluoride) thin coating for dendrite-free and high-performance Lithium metal anodes. Advanced Energy Materials. 2018;**8**:1701482

[50] He Y, Xu H, Shi J, Liu P, Tian Z, Dong N, et al. Polydopamine coating layer modified current collector for dendrite-free Li metal anode. Energy Storage Materials. 2019;**23**:418-426

[51] Zhu B, Jin Y, Hu X, Zheng Q, Zhang S, Wang Q, et al. Poly(dimethylsiloxane) thin film as a stable interfacial layer for high-performance Lithium-metal battery anodes. Advanced Materials. 2017;**29**:1603755

[52] Wen S, Li Z, Zou C, Zhong W, Wang C, Chen J, et al. Improved performances of lithium-ion batteries by conductive polymer modified copper current collector. New Journal of Chemistry. 2021;**45**:10541-10548